NORTH CAROLINA
STATE BOARD OF COMMUNITY COLLEGES
LIBRARIES
ASHEVILLE-BUNCOMBE TECHNICAL COMMUNITY COLLEGE

Y0-EGA-423

DISCARDED

JUN 1 2 2025

Safety and Environmental Training

*We bring more than a paycheck
to our loved ones and family
We bring asbestosis, silicosis,
brown lung, black lung disease
radiation that hits the children
before they've really been conceived.*

*I want more pay
but all I got today
more than I bargained for
when I walked through that door.*

*Workers lend an ear,
its important that you know.
With every job there is fear,
that disease will take its toll.
If not disease then injury,
may befall your lot.
If not injury then stress
is going to tie you up in knots...*

                    From "More Than a Paycheck"
                    Lyrics and Music by Ysaye M. Barnwell,
                    Barnwell's Notes Co.
                    Copyright 1982

Safety and Environmental Training

Using Compliance To
Improve Your Company

Dawn A. Baldwin

VAN NOSTRAND REINHOLD
New York

Copyright © 1992 by Van Nostrand Reinhold

Library of Congress Catalog Card Number 92-15212
ISBN 0-442-01066-4

All rights reserved. No part of this work covered by
the copyright hereon may be reproduced or used in any
form or by any means—graphic, electronic, or
mechanical, including photocopying, recording, taping,
or information storage and retrieval systems—without
written permission of the publisher.

Printed in the United States of America

Van Nostrand Reinhold
115 Fifth Avenue
New York, New York 10003

Chapman and Hall
2-6 Boundary Row
London, SE 1 8HN, England

Thomas Nelson Australia
102 Dodds Street
South Melbourne 3205
Victoria, Australia

Nelson Canada
1120 Birchmount Road
Scarborough, Ontario M1K 5G4, Canada

16 15 14 13 12 11 10 9 8 7 6 5 4 3 2 1

Library of Congress Cataloging-in-Publication Data

Baldwin, Dawn A., 1966-
    Safety and environmental training : using compliance to improve
your company / Dawn A. Baldwin.
      p.   cm.
    Includes bibliographical references and index.
    ISBN 0-442-01066-4
    1. Industrial safety.   2. Industrial hygiene.   I. Title.
T55.B25   1992
658.4′08—dc20
                                                    92-15212
                                                      CIP

Contents

Preface   ix

1 Compliance as Opportunity   1
    LEARNING TO LOVE REGULATIONS   5
    JUST SOME OF THE BENEFITS   8

2 The Elements of Training   16
    CONSISTENCY   16
    COMMUNICATION   20
    GOALS   21
    DISCIPLINE   23
    LIFE   25
    LEADERSHIP   26

3 The Right-to-Know   28
    POWERFUL POSSIBILITIES   29
    THE NEED TO KNOW   35
    DECIPHERING THE REGULATORY CODE   40
    THE NETWORK   58

4 Tools of the Trade   63
    WHAT DO OSHA'S PERSONAL PROTECTIVE EQUIPMENT STANDARDS REQUIRE?   64
    GLASSES AND GOGGLES   67
    RESPIRATORS   72

5 Plan, Prepare, Prevent   83
    EMPLOYEE EMERGENCY PLANS   86
    FIRE PREVENTION PLANS   103

6 More Than Maintenance 109
THE LOCKOUT/TAGOUT RULE   111
USING STANDARD OPERATING PROCEDURES   137

7 Before It Breaks, Fix It! 140
DEFINING THE SAFE OPERATING PROCEDURE   141
SET GOALS, GET RESULTS   143
PRIORITY PROCEDURES   148

8 Combine and Conquer 161
THE MANAGEMENT CONNECTION   161
COMBINING TRAINING   172
REDEFINING THE SAFETY MEETING   178

9 A Little Respect 182
GIVE CREDIT WHERE CREDIT IS DUE   184
REMIND THEM WHAT THEY ALREADY KNOW   187
UNCOMMON COMMON SENSE   190
THE POWER OF LISTENING   192
CHANGE, CHANGE, CHANGE   195
MAKE IT MATTER   197
KEEP IT SIMPLE   199
WHOSE LIFE IS IT, ANYWAY?   202

Postcript 206

Appendix 209
ASSOCIATIONS   211
SOURCE BOOKS AND ASSISTANCE GUIDES   212
ASSISTANCE ORGANIZATIONS   215
TRAINING SCHOOLS (TRAIN-THE-TRAINER)   215
TOXICOLOGY AND EXPOSURE LIMITS   218
PERIODICALS   218

Glossary 220
OSHA HEALTH HAZARDS   220
OSHA PHYSICAL HAZARDS   221
WORDS ON MATERIAL SAFETY DATA SHEETS   222

Index 225

To Janet and Rozelle

Preface

While writing this book, I have given a lot of thought to the term "wellness." Over the past year, wellness programs or health promotion programs have been difficult to ignore. In publications as varied as *Professional Safety* and *American Psychologist*, people are writing about them. Even the federal government is in on wellness, having published a lengthy book in 1991 titled *Healthy People 2000: National Health Promotion and Disease Prevention Objectives* that is all about the development and implementation of wellness programs.

But after considerable thinking and reading about wellness, I'm still asking myself: *what is it?* What does it mean to *be well?*

Most health promotion programs define wellness as personal health, as the absence of illness or injury. According to these programs, being well means developing good health habits involving exercise and nutrition, giving up smoking, ending substance abuse, and adopting safety habits such as wearing seat belts. As the journal articles report, more and more cost-conscious employers and insurers are realizing the merits, at many levels, of promoting personal wellness to their respective employees and policyholders.

Is this all there is to it? We modify our health habits and lifestyles through exercise and dietary regimens, stop smoking and drinking, and start wearing seat belts, and at some future date, we become *well*. I don't mean to imply that doing all of these things is easy, it's not. I'm just wondering if wellness really begins and ends with a person's skin. What about the role of our physical, social and emotional *environments?* In other words, what does our *well-being* have to do with wellness?

When I think of well-being, I think of *belonging,* having a strong sense of community and of commitment to those who share my physical or social situation. Since I am a physical creature, my feelings of well-being are, to a great extent, dependent on the quality of the environment in which I find myself. If the environment in which I work aids in protecting me from injury and illness, then I will develop a sense of well-being at work. Further, if I feel I am a part of my work environment to the extent that my presence is uniquely valued and my ideas affirmed, my well-being will be enhanced.

Likewise, in order to have a sense of well-being in the larger environment in which I live, several basic life-nurturing conditions must exist: I must feel no threat from crime or violence; I must have unpolluted air to breathe and water to drink; I must have sufficient food to sustain me; and I must have a balance of social interaction and privacy. If these conditions exist, my well-being is possible.

And well-being has wonderful side effects, improved personal health being just one of them. People with a strong sense of well-being are proactive about all aspects of their lives. Feeling a real connection to those with whom they share their lives at work and away from work, these people are very much aware of the interdependence of all social groups, regardless of size. Seeing themselves as members of many such groups, including family, work, community, country, and world, people with a healthy sense of well-being feel a responsibility to be the best they can be. Why? Because they know that *who they are* and *what they do* is very important to the well-being of many, many other people.

Is personal health or wellness *sustainable* for people without a sense of well-being? I really don't think it is. The influence of environment cannot be ignored forever.

What does any of this have to do with safety and environmental compliance? Safety and environmental regulations focus your attention on the workplace conditions that have the greatest impact on employee well-being: health and physical hazards, pollution and waste, the flow of information from management to employees, and the presence (or absence) of an atmosphere of openness, trust, and respect. If well-being is a product of our physical, social and emotional environments, can you imagine a better way to begin creating this feeling in your employees than through training in safety and environmental health? I can't.

As someone responsible for safety and environmental compliance, you have the opportunity to influence the *physical* environment at your workplace by providing safety devices and personal protective equipment, by implementing procedures for safe hazardous materials handling and waste minimization, and by developing methods for preventing spills and fires and responding to emergencies. You also have the opportunity to shape the *social* environment at your workplace by including everyone in the development of safety and environmental policies, and by breaking down the barriers of power and position that so often conspire to keep management and employees from working together toward shared goals. Finally, you have the opportunity to enhance the *emotional* environment at your workplace by valuing the uniqueness of each employee, by trusting and being trustworthy, and by communicating openly, which means pausing to listen.

If your workplace environment was effected on these three levels, would it be recognizable as the same organization it is today? Can you imagine the long range impacts, in terms of accidents, health care, production capability, job turnover and quality? Did you realize safety and environmental compliance held so much power?

I believe that expertise is only valuable in relation to our ability to share it. And what is expertise but a *collection of unique experiences?* This book is an account of my experiences in safety and environmental training. All the examples, the case histories, and the vignettes are real. The interpretations of the regulations are not based on theory or supposition. They have been arrived at through trial and error. They are included here because I have seen them fail or watched them work.

In this book I take a very positive view of regulations. I present regulations not as arbitrary rules, but as guidance documents. I take this approach not because I think all safety and environmental regulations make sense, or because I like the way they're written, or because I think all regulators are reasonable people, but because I see it as the only viable option. I'm not satisfied with survival. I want *sustainable improvement.*

I also happen to believe in the self-fulfilling prophesy. If I expect regulatory compliance to be an absurd, ineffectual, negative experience, it will be. And I, for one, can't afford that kind of waste. If, however, I expect to figure out a way to use regulatory compliance to my benefit so that it becomes a logical, effective, positive force in the workplace, then I have a chance to achieve the sustainable improvements that are my goal.

Acknowledgments

I would like to thank the many reviewers whose comments have enhanced the quality of this book. Thanks also to Kevin Kamperman at Georgia Tech Research Institute, Neal Lorenzi at *Professional Safety,* the Materials Technicians at American Resource Recovery Corporation, all the clients of Wimmer Baldwin Associates, Inc. who have given me the chance to practice what I preach, Danielle Rein, whose attention to detail has made every word matter, and Hilary Jacobs without whose diligence and faith this road would have been much harder indeed. And Milton, since thanks hardly suffice, I acknowledge all you've done for love.

Safety and Environmental Training

1
Compliance as Opportunity

*We are governed not by
armies and police
but by ideas.*

Mona Caird

The problem with the perspective of many modern employers and managers is that we spend too much time looking down and backwards: down at the bottom line, and backwards to what worked in the past. As many are learning—some quite painfully—what worked in the past with regard to employee safety training and environmental compliance can turn today's bottom line bright red.

There was a time when employers could get by on the premise that ignorance is bliss, that what neither we nor our employees knew about the hazards of the workplace, or *chose to admit,* couldn't hurt us. But that was when safety concerns were limited to monitoring the physical hazards of the workplace: slippery walkways, exposed machinery, blocked aisles. That was before any connection was made between the way an employee does his or her job today and the impact those actions can have on personal health and the environment in years to come.

Times have changed. Now, thanks in great part to the OSHA Hazard Communication Standard and the Resource Conservation and Recovery Act, attention to the connection between today's work behaviors and tomorrow's health effects and environmental impacts has been legislated. You have to be able to account for the ways in which your workplace impacts human health and the environment. You have to be able to account for the steps you have taken to train your employees to minimize the impact of their work activities on human health and the environment. Your responsibility for *today* extends far into a future you cannot see or predict.

You didn't ask for this responsibility. Nevertheless, it's not optional. Rather than bliss, ignorance of the hazards of your workplace has become a liability no manager can afford. With the promulgation of OSHA and RCRA regulations, *safety* has been transformed into an umbrella term incorporating environmental, life, and job quality issues as part and parcel of the same subject. Today, safety training is as much concerned with the *future impacts* of work behaviors on health and wellness as it is with how those behaviors impact the employee (and the company) in the present tense. In this new focus on both the short- and long-term effects of work behaviors, safety training now must address a range of topics unheard of a decade ago. Hazardous materials and hazardous waste management, emergency response, spill control, waste minimization, and pollution prevention are now synonymous with the term *safety*.

In the past the authoritarian "do as I say, not as I do" approach to safety may have been the standard. Now, with safety's new long-range focus, an approach of such dubious effectiveness is dangerous at best. Why? Because it is crucial that your training develop employees who *thinkingly* adhere to certain behaviors because those behaviors *make sense*. Old style threats and coercion only serve to create robots who do one thing when you are watching and another when your back is turned. In your highly regulated workplace, where you must be accountable for your employees' actions far into the future, it is essential that their behavior is consistent.

Clearly, if it is to effect change, safety training cannot be authoritarian. Neither can it be limited by the regulations that govern it. Instead, it must address what employees must do to protect human health and the environment at your workplace, while encouraging their independence and creative thought. Today, safety training must address your employees' attitudes toward personal, family, and community wellness.

Looking back to what worked in the past, compliance with training requirements seems a four-lettered burden. It doesn't have to be. It can be a four-star opportunity to invest in the future, depending upon your vision and where you choose to aim it.

Theirs not to reason why;
Theirs but to do and die.

>from *The Charge of The Light Brigade*
>Alfred Tennyson

Just for a moment, ask yourself *why*.

Why do you have a training program? Why do you pull your employees off the job to talk about hazardous materials or waste or to fit them for a respirator? Why do you attend all of those training courses and seminars? Why do you spend money on regulatory updates, training program manuals, periodicals about training, training consultants and videos? What's the *point?*

What would you say? Perhaps: *to comply with the law, of course.*

But training regulations don't exist simply for the purpose of being complied with. Compliance is not supposed to be its own reward. Compliance is supposed to serve a *purpose*.

So what's the point of the training itself? What is it supposed to accomplish? What purpose is it supposed to serve? What are you supposed to get out of it?

You say: *I'm supposed to stay out of court and jail, that's what I'm supposed to get out of it. What do I care if it has a purpose? It's a law. I have to do what it says whether it has a point or a purpose or I like it or don't. Period.*

Most of us figure it's not our place to ask *why* where regulations are concerned. A training regulation is a law. And by that definition we know everything we need to know about it before we read word one: We know that laws, generally speaking, are things we would rather not obey. We know that laws butt into our business and interrupt whatever we're enjoying, only to replace it with something we won't enjoy at all. We know they tend to make life difficult or uncomfortable. We know it's often hard to pinpoint exactly what a law is good for. If, in fact, they have any benefits for us at all, it's often very difficult to find them.

But despite all of this, we obey laws because we're scared not to. If we don't obey, we might get caught. Then we'll have to pay fines, our name will be in the news, or we may even have to go to jail. Whether or not a law is reasonable, fair, or makes any sense at all has no bearing whatsoever on whether or not we have to obey it. Anyone who has argued with a police officer over a speeding ticket knows this. So we learn to buckle under and obey the law "because it's a law and we'll be punished if we don't." We've learned that questioning why is pointless and only gets us into trouble.

Actually, this attitude toward laws begins when we're children, when the makers of the laws we live by are not government agencies but parents. Why do I have to go to bed? Why do I have to eat my carrots? Why can't I have any more cookies? Why can't I stay out late? Why can't I go to this party?

And the law answers: *Because I said so.*

And so, to avoid being sent to our room, having our allowance docked, or being grounded, we buckle under. In the face of the law, asking why never gets us anything, except punishment. Laws are nonnegotiable. Training regulations are different.

When a police officer pulls you over for going 45 mph in a 35 mph zone, he doesn't give you a chance to explain why you think 45 mph is a sufficiently safe speed to protect your life and health and the life and health of others in that particular neighborhood. When you file your tax return, Uncle Sam doesn't ask you to document why you think your money would be better spent by you than by him. With those laws there is little, if any, room for interpretation, for reasoning.

However, with training regulations not only is there room for interpretation and reasoning, it's *absolutely required.* Inherent in compliance is a process of negotiation and interpretation, of analysis. Your thought and input is essential.

That's the catch. *Mindless obedience to arbitrary rules won't work here.*

Assume for a moment that a direct mandate of the regulations is that whatever you do to meet their requirements, *it must make sense.* Different, huh?

Sure, sometimes the language of the regulations is confusing. Sure, sometimes the regulations read like they've been written by someone who has never darkened the door of an industrial operation. Put all that aside for a moment. Open your mind. Think of safety and environmental regulations as something that demands that everything you do to comply with them must be *appropriate to your workers, must apply to your operation,* and must actually *improve things at your workplace.*

Powerful possibilities are revealed through this approach. Safety and environmental training *has* to be good for the company, *has* to have some tangible benefit, or its not worth doing. Can this be true? Are the regulations really saying: If it doesn't apply, don't do it; if it's a waste of time, change it; if it doesn't make sense, don't bother? Does this mean that the only way to judge your compliance is by the degree of positive impact it is having at your company or by the degree to which it is *not* a boring, tedious, ineffectual, impractical, burdensome waste of time and money?

Hmmm.

This is, of course, not the customary attitude. Most of the regulated masses assume the regulations won't make sense. We knuckle under and grind our teeth, *expecting* to do inappropriate, impractical, nonsensical things to comply. We *expect* not to understand what we're doing or why we're doing it. And we *expect* that our compliance efforts will have no impact at our company and no benefits for us. This attitude is a self-fulfilling prophesy. The more mindless and arbitrary our compliance activities, the more ineffectual our written programs and training classes, the more comfortable we are. The status quo remains unthreatened while we the regulated can complain: "Just look at what OSHA makes me do. What a waste!"

Training regulations assume a level of sophistication and personal involvement on the part of employers that is totally absent in laws regarding parking or smoking in airplanes. Those laws demand you throw up your hands and obey, whether or not you think obedience makes sense. Training regulations offer you the opportunity to move beyond the nonsensical. Instead of taking them at face value, you have the opportunity to cut them apart, distill them, mold and shape them into something that is useful for you and your company. To transform them from hindrance to assistance. Instead of submitting to them as arbitrary rules or laws, think of training regulations as guidelines you follow to help you develop safety policies and procedures that meet specific needs at your workplace.

Your whole compliance effort must be geared toward meeting those needs. Meeting the needs you have identified must become the motivation behind every written program, labeling system, inspection, training class, and safety meeting.

Unless you know why you're complying, it's virtually impossible to comply with the requirements of a training regulation. Unless you know why you're complying, you won't be able to identify where your compliance should begin, and you won't recognize the end if and when you come to it.

One word of warning: Mindless, grudging submission to arbitrary laws is a whole heck of a lot easier than thoughtful compliance for the purpose of making positive change. It's a lot easier to preserve the status quo than to renovate it.

LEARNING TO LOVE REGULATIONS

Training regulations are designed as *guidelines* for meeting specific safety, health, and environmental management *needs*. This is an important concept that is often overlooked. But, once you grasp it, you'll never hold a training requirement at arm's length again.

What is a guideline? It is a standard by which to make a judgement or determine a policy or course of action. There are two key characteristics of a guideline:

1. It is not an end in and of itself; it is a means to one or more ends.
2. It requires the active interest and involvement of the person or persons required to follow it.

What is a need? It is a lack or deficiency. If someone needs something, there is an implied *or else* following the statement of need. There is an implied or predicted negative result that will occur if the need is not met. If safety needs are not met, accidents and injuries will occur. If environmental management needs are not met, environmental impacts will occur.

Where known safety, health, or environmental management *deficiencies* exist, training regulations provide employers *standards by which they must make policies or determine courses of action for correcting those deficiencies* at their particular workplace. Yes, the law says you *must* take action. You must make a policy about particular safety, health, and environmental issues that impact your workplace. You must establish a plan for meeting the particular safety, health, and environmental needs that exist.

But what sets training regulations apart is the fact that they are, by definition, performance-oriented. This means that each regulation addresses a specific workplace need and challenges employers to figure out and document a method for meeting the need that will work, or *perform,* for them. Training regulations put the onus on the employer to use their requirements to make their company a safer, healthier place to work. Training regulations put the ball in your court. This allows you great freedom and flexibility in meeting training requirements.

Training regulations indicate only general categories of information that should be covered in your program, not how the information relates to your site or how it should be presented to your employees. For example, Right-to-Know training is supposed to include the methods and observations that employees can use to detect a release of a hazardous chemical in their work area. The way in which this is handled in a training class will vary greatly from site to site, depending upon the types and amounts of hazardous materials present, the employee's work activity, established standard operating procedures, required protective equipment, and available engineering controls and monitoring devices. Similarly, both the Hazard Communication Standard and the Lockout/Tagout Rule allow employers to design hazard warning labels that fit their workplace operations and to train their employees in whatever system they decide will perform best.

In order to meet these requirements, the employer must look at the law and ask *why*. Why do my employees need to know when a hazardous material has been released? Why do my employees need labels on containers of hazardous materials? Only when you know why you must comply (read: why there is a need) can you go about the business of complying (read: filling that need).

Most regulations that require training don't even say how long training should last or how often it should be conducted. Why? Because it's up to the employer to figure out how much training it's going to take to meet the needs he or she has identified.

If training regulations could defend themselves, what would they tell employers? Perhaps they would say something like this: *My purpose is not to make unreasonable demands. I want you to use me to set policies and courses of action that are practical, that will perform for you and your employees. Use me as a tool to correct your deficiencies now, so that you can stop worrying about what might happen later if you continue to let those needs fester and grow. Follow my lead. Don't gamble with the lives of your employees, with the finances and reputation of your company, with your own future. I am the alternative to crisis management. I am the easier way.*

Let's look at five "popular" training regulations and see if this defense holds true. Are these regulations rigid laws or performance-oriented guidelines for meeting real workplace needs?

Example #1: Why Right-To-Know Training?

Over time, employees have shown again and again that, when they work with hazardous materials, systems need to be in place for informing them of the hazards of those materials. If they aren't informed, if they don't know how to handle the materials safely, how to protect themselves from overexposure, or how to recognize if they are overexposed, one thing is certain: Sooner or later, someone gets hurt. History has shown that the severity of the injury can vary from a skin rash to chronic disease to sudden death. On the flip side, employees who are uninformed of the hazards of the chemicals they use can cause property damage, ranging from destruction of process equipment to chemical explosions or fires, simply because they didn't know any better.

The Hazard Communication Standard exists to provide employers with a guideline for developing an educational system that will meet this need at their workplace. Therefore, *no one* has to get hurt and property *doesn't* have to be damaged before employees get a chance to learn how chemicals behave.

Example #2: Why Lockout/Tagout Training?

Repeatedly, employees have demonstrated that, when energized equipment is down for repairs, they need a system for letting each other know that the equipment is being worked on and who's working on it. If they don't have such a system, it is inevitable that one day someone will walk by a switch or lever or circuit breaker in the "off" position and flip it "on"—at the wrong time. When that happens, employees lose fingers and legs in moving equipment. They are maimed, burned, electrocuted, or even killed.

The Lockout/Tagout Rule provides employers with a guideline for developing a system for meeting this need at their workplace. You say you've been in business for 30 years and never had anyone get hurt? The Lockout/Tagout Rule says it would be a shame to ruin such a wonderful record with an accident. If you've gone 30 years with no injuries, you're probably due for a little bad luck. Why not talk about this subject with your employees and shift the odds back in your favor? What (or who) can you afford to lose?

Example #3: Why Respirator Training?

Employees everywhere have proven that, if you give somebody a respirator and don't tell him or her why it is worn, when it is worn, or how it is worn, one of two things will happen: the respirator will never be worn, or it will be worn improperly. And if he or she doesn't know how to wear it, it is unlikely he or she will store, clean, or replace it properly either. If one or more employees work in an area where exposure levels are such that a respirator is required only occasionally or all of the time, and they don't wear the respirator or they wear it improperly (which is basically the same thing), then their employer is responsible for allowing them to endanger their health—allowing them to run the risk of chronic health effects, nervous system and organ damage, and disease. In order to protect themselves (and thereby protect their employer), those employees need information about their respirator and guidance in why, when, and how to wear it.

The Respiratory Protection Standard provides employers with a guideline for making sure employees who are supposed to wear respirators actually wear them and wear them properly.

Example #4: Why RCRA Training?

Records indicate that, if generators of hazardous waste do not have a method or procedure for informing their employees of how to store, handle, label, and prepare the waste for transport or respond to spills, leaks, or fires involving the waste, then none of these tasks will get done to the EPA's liking. Employees will run the risk of causing harm to themselves, to others, or to the environment. And the generator will be in big trouble, risking harm to his or her employees, harm to the environment that can be traced back to his or her negligence, fines, lawsuits, bad press, and even imprisonment.

That's why the RCRA regulations governing treatment, storage, and disposal facilities, and generators include personnel training requirements: to help employers help themselves. These requirements provide employers with a guideline of what their employees need to know in order to manage hazardous waste in an environmentally sound and safe manner.

Example #5: Why Hazardous Waste Operations and Emergency Response Training?

Experience has proven that employees who work at Superfund sites or other government-controlled hazardous waste clean-ups or treatment storage and disposal facilities, or who serve on emergency response teams anywhere, need specialized training and information in order to avoid hurting themselves, other people, or the environment in the process of doing their job. What employees need to know will vary greatly depending on what they are required to do and under whose jurisdiction they are doing it. Workers at government-controlled clean-ups need site-specific information, or else they may very well do something that is not sanctioned by the agency with jurisdiction over the project. Workers at treatment, storage, and disposal facilities need information that is integrated into their standard operating procedures, otherwise they risk harming themselves, others, or the environment through lack of appropriate knowledge.

Emergency response workers need information geared to a specific response assignment that is general enough to be applied to a variety of incidents so that they don't freeze up or panic if the emergency they encounter this week isn't identical to last week's.

The Hazardous Waste Operations and Emergency Response Standard provides the specific types of employers named in the standard with detailed guidelines for meeting the educational needs of their employees.

JUST SOME OF THE BENEFITS

Why do training regulations exist? Each training regulation exists as a guideline to help you meet a real need at your company.

You see, training regulations are not arbitrary. They don't exist just to give regulators something to do. Their purpose is not to pick on you or waste your time and money. They aren't just another way for the government to generate funds. Believe it or not, they don't exist for anybody's benefit but *yours*.

Compliance as Opportunity   9

If a training regulation could speak in its own defense, what else would it say? Perhaps it would tell employers: *First, before you do anything, read me and ask yourself why I do or don't apply to you. Ask yourself what you risk losing if you don't follow my lead. How happy are you with the status quo? Ask yourself just exactly how much you can afford not to gain.*

Ask yourself: *Do I have hazardous materials as defined by the Right-to-Know law? Or energized equipment as defined by the Lockout/Tagout Rule? Do my employees need to wear respirators? Do I generate hazardous waste, as defined by Title 40 CFR Part 261? Or am I an operation as described by the standard for Hazardous Waste Operations and Emergency Response?*

If your answer to any of these questions is yes, then ask yourself: *Why do my employees need training and information about these materials, equipment, processes, or procedures?*

As established above, employees need training and information in order to avoid the predictable negative outcomes of being uninformed and uneducated. So they won't get burned, maimed or become ill. So they won't cause chemical reactions, explosions, or fires. So they won't harm others or the environment. So they won't, out of ignorance, disobey the laws their employer is governed by.

That's the obvious answer. And, ideally, that answer alone would be enough to convince every employer to embrace training regulations as the single most powerful tool they can use to protect the future of their company.

Often, however, it's just not enough. Why? Because employers don't perceive that they have any deficiencies in any of these areas. They say they don't have anybody getting maimed or killed, blowing the place up, getting an arm caught in equipment, mismanaging waste, or contaminating the environment. They say their employees are smart, experienced, and hardworking. They know what they're doing. The last thing they want or need to do is waste time sitting around in a bunch of training classes.

If that is your response, then one of two things is true:

1. Your employees really are smart and experienced and already know everything they need to know about hazardous materials, hazardous waste, respirators, energized equipment, environmental regulations, and emergency response in order to protect themselves and you. In this case you don't need to comply with training regulations. You need to document how your employees came by this knowledge and send it to OSHA.
2. You are living on borrowed time. Your good luck has lulled you into a false sense of security. And you probably don't know as much about what goes on in your workplace as you think you do.

Either way, if the idea of looking at your workplace or your employees as needy or lacking just doesn't settle well in your stomach, you might ask one last question before you decide whether or not it's worthwhile for you to get interested in training: *What are some of the benefits of having well-informed employees?*

Let's imagine that we have a business where accidents, injuries, and illnesses are impossible. There is a special force field that exists around our workplace that won't allow such things to happen. Therefore, eliminating accidents, injuries, and illness cannot be a benefit of our training. We have training and information programs for *other* reasons. What might those be?

Improving or Streamlining Production

Employees who regularly attend training classes are more self-confident than employees who don't. As the classes encourage them to examine their work environment and their personal work habits with respect to safety and environmental management, they begin to voice ideas and suggestions for improving the workplace. Employees who feel they have no voice will keep their mouths closed for years, perhaps even for their entire work lives, knowing a better way, but feeling too powerless under management to do anything about it. This is an immeasurable tragedy, because employees who produce products, operate equipment, and handle chemicals know more about their job and *how it can be done better* than every manager at your company plus a roomful of engineers. Employers who tap into their employees' knowledge by encouraging them to talk about their ideas will, invariably, produce a better product, provide better service, and just generally create a better company.

Case in Point: Materials handlers at a hazardous waste management company suffer for well over a year with a key piece of process equipment that they know is designed wrong. First, the shaft on the blender into which they dump barrels of hazardous waste is too short. This allows solids to fall out, which, in turn, requires one of them to get down in the blender and shovel it out once a day. Second, the trip bar on the ramp where they stand to dump drums is too far away from the blender opening. This means that every time they dump a drum, waste splashes everywhere, much worse than it would if the trip bar was closer to the blender. But they don't tell anyone, because they don't think it's their "place" to tell management they goofed. Finally, several months after an on-going OSHA/RCRA training program is implemented at the facility, the men speak up. The training session that day is on limiting exposure to chemical vapors by reducing them at their source. One of the men says: "Well, I may not know much, but I've never understood why they made that shaft so short. We could cut the vapors in half if we didn't have to leave the lid up and get down in there ourselves to dig the thing out." Soon all of the men are talking about problems with the blending equipment. Within a month, the improvements are made.

Boosting Morale and Encouraging Job Pride

Employees who have input into how they do their job care more about it. The input does not have to be earth-shaking or complex, as long as it results in some discernible *change in the workplace*. In fact, it's often the little "pebble-in-your-shoe" problems that managers don't or can't see that employees will silently suffer with for years unless encouraged to speak out. Again, this is a tragedy because, most often, these problems would not be expensive to fix if addressed in a timely fashion. However, left to fester, they end up costing the employee and the company a bundle in terms of safety, job performance, production quality, and turnover. The bottom line is that a suffering employee is not happy. An employee who has no evidence that management cares about him or her as a person or is curious enough to ask him or her what it's like to do his or her job will begin to feel exploited. If this resentment is allowed to grow, the employee can become aggressively non-caring about the quality of the work, as well as a danger to him or herself, co-workers, and the company as a whole. In contrast, an employee who knows that management views him or her as an individual with a voice will feel he or she has some autonomy over his or her worklife. Instead of feeling exploited, the employee will feel proud and be happier.

Case in Point: The maintenance employees at a plating company figure that getting holes in their jeans, shirts, and shoes from caustic or acid spray is just part of the job. No one has ever asked them about it. The word around the plant is that, number one, the boss is too cheap to buy any protective clothing, and, number two, he doesn't like sissy complainers. They think that if the boss had to be out here crawling around heated vats of corrosives, he'd change his tune. But then they're just maintenance men, and the boss doesn't give a damn what they think. In really low moments they almost wish one of them would get hurt so that the boss would have to pay, and pay big. But when the company hires an outside consultant to conduct Right-To-Know training, the men open up and express their concerns. The consultant carries these concerns to management as part of her compliance recommendations. The company establishes a policy regarding personal protective equipment. Everybody wins.

Increasing Awareness of Environmental Impacts and Regulatory Controls

Some of the environmental regulations you must comply with do not directly mandate employee training. This does not mean, however, that you should feel restricted from letting your employees know about them or that the compliance process would not be improved if employees were more aware of regulations and regulators as a whole. In fact, the safety and environmental regulations that do require training offer an excellent avenue for getting employees involved in the *total compliance process*. If employers don't tell their employees, first, that the company is regulated, and second, what the company, through each of them, does to comply with those regulations, how else are they going to know it?

The average man or woman on the street does not know how and why business and industry is regulated. They don't know that companies such as yours must meet a variety of regulatory requirements, regarding everything from how and where hazardous materials are stored and hazardous wastes are labeled to the number of pounds of *volatile organic compounds* (VOC) that are being emitted into the air and the constituents in storm water run-off. You wouldn't pull a person at random off the street and put him in charge of your life's savings without any input or guidance from you. Of course not. That would be crazy. Well, it's just as crazy to allow your employees to operate your business ignorant of how their actions impact the environment (inside and outside the company walls) and thus your regulatory compliance status. This is not to suggest that you use doom and gloom scare tactics to coerce your employees into toeing the line. It is to suggest, however, that employees who understand that their job is *important enough to be regulated,* and that, depending on how they do their job, it can have a *negative or positive impact on their environment* inside and outside work are going to be interested in their work and take ore care and pride in doing it.

This team approach to compliance can have several benefits for you. First, you don't have to worry so much about employees "blowing the whistle" on some part of the operation that they are in the dark about. Whether or not you are handling your wastewater monitoring or your waste oil disposal correctly has no bearing on the fact that, if a disgruntled employee is uninformed about some aspect of the plant's operation, especially if it has to do with the boogeyman "waste," he may decide to place a few phone calls in an attempt to stir up trouble. Whatever the outcome, the experience will cast you in a suspicious light. If an employee is aware of your compliance obligations and what you and he or she together are doing to meet those obligations, he or she is less likely to do something like this.

Second, you're simply more likely to stay in compliance if everyone is involved in the compliance process. If you required your employees to remove the blinders that confine them to one isolated job and to see the company as an interdependent unit, where each action impacts everyone and where everyone is responsible not only to the boss, but to each other, how might your operation improve? The examples are endless.

Case in Point: You are a plater with a wastewater treatment system that must be monitored once a week for pH. You could have your process chemist go out to the plating line once a week and take a sample while your line workers wonder what he's doing. Or, as part of their Right-to-Know training, you could hold a meeting with your line workers and let them know what pH means and why the pH of the wastewater must stay within a certain range. You could tell them that, since the water goes to the city treatment plant, the city requires you to file quarterly reports indicating the results of the testing. You could tell them how important it is that they handle chemicals carefully and minimize spills along the plating line, because too much acid or caustic causes a sudden rise or fall in the pH, which is what the city gets upset about. You could even have the line foreman start monitoring the pH himself.

Case in Point: You are a large-quantity generator of waste paint-related material. You have one man in shipping and receiving who has been trained to fill out labels and manifests and schedule waste pick-ups. But there is a problem. Month after month the treatment, storage, and disposal facility where you send the waste keeps charging you extra because there is always trash (paper cups, cans, rags) mixed in with the waste. You've sent memos to your hazardous waste man in shipping and receiving, but the problem keeps recurring. What would happen if, as part of your RCRA Hazardous Waste Generator compliance, you held a meeting with every employee involved in generating hazardous waste and told them why it is so important to keep regular trash out of the hazardous waste drums? You might discover that your employees had no idea that the hazardous waste was being sent somewhere to be recycled or blended into fuel. You might discover that they just thought hazardous waste was waste; it was going to be thrown away somewhere. And so a few potato chip bags or old gloves mixed in wouldn't matter at all.

Minimizing Waste

One of the key indicators of how efficiently a company is operated can be found in the computation of the amount of waste generated per pound of saleable material produced. How much does it cost you to generate waste—not just the disposal fees, but the cost of the paperwork, management time, labor, and even the containers the waste is stored in? Often, waste is generated because equipment is malfunctioning, because it hasn't been repaired properly or serviced regularly, because it is being run too hot or too fast, or because employees are using too little or too much of a certain lubricating fluid or other raw material. How much does that cost you in down time and repairs?

Maintenance and housekeeping procedures are essential to effective waste minimization. If you want to find out if you are wasting your hazardous materials inventory through improper storage or use of those materials, walk around your workplace and observe housekeeping procedures. If you want to know how much inventory you are wasting through leaking hoses, valves, and gaskets, or improperly adjusted machines, watch your equipment operators or maintenance crew at work.

Waste minimization will be a benefit of any safety or environmental training that you conduct because procedures that reduce waste are also those that are safest for your employees and the environment inside and outside your company. They are also the most economical for you.

Case in Point: You are a service director at a car dealership. In order to cut down on hazardous waste, you have recently done two things in your paint shop. First, you bought two closed loop gun flushers. These units recycle the thinner that is used to clean paint guns, leaving a very small amount of sludge to be disposed of. Second, you switched to a paint system that is more streamlined. It requires one less step in the application process than the previous system, meaning one less product to use and one less time during every job that the guns have to be cleaned. These were management decisions, made to save money. But their impact doesn't have to end there. The new paint system and gun flushers might be discussed during Right-to-Know training as examples of efforts the company is making to reduce employee exposure to hazardous materials. Or they could be presented during RCRA Hazardous Waste training as part of the company's effort to reduce waste in preparation for waste minimization requirements.

Case in Point: You are a furniture manufacturer with the responsibility (under the Superfund Amendments and Reauthorization Act (SARA) Title III, Section 313) of tracking what happens to thousands of gallons of listed hazardous materials over the course of a year. On what is called a "Form R," you must indicate how much is used in process, how much is disposed of as hazardous waste, and how much evaporates into the atmosphere. You also have to maintain several air pollution permits for VOC emissions from specific process equipment. And you are looking ahead to having to have a Waste Minimization Plan to decrease your hazardous waste generation by 25% and a Pollution Prevention Plan to control any pollutants from your facility that might enter air or water. Your production workers may not be able to recognize a Form R from an Air Permit, and they don't need to. What they *do* need to be aware of, however, is that their company's impact on the surrounding community, the community they inhabit, is monitored. And that the standards by which it is monitored are getting stricter every year. Employees need to know that how they manage hazardous materials and hazardous wastes in their work effects their environment *outside work.* Every time they leave tops off of containers of volatile solvents or neglect to clean-up spills, they are adding to the air pollution, not only of their personal work environment, but of the environment their children live in. Right-to-Know, Respiratory Protection, and Hazardous Waste training all provide opportunity to bring this up, and all will be made more meaningful by its mention. If you can convince your employees to make a personal commitment to air pollution prevention, your compliance with applicable regulations will be made easier, you will waste less chemicals through evaporation, and your facility will be safer. In addition, your employees will feel better physically and they will feel good about their work emotionally. They will be less cynical and less apathetic.

Why is it easy to love training regulations? Because they have your best interests at heart. They encourage you to be selfish, to ask: *What can I get out of this?* And they guarantee that, if you train with the above four categories of benefits in mind, you actually will succeed in erecting a kind of force field around your company, where injuries, occupational illness, pollution, wastefulness, and uncaring are not tolerated. The protection of safety and health actually will be just one of the many results of the training, the purpose of which is not just the improvement of the workplace and the worklife, but life *beyond* work as well.

2
The Elements of Training

*Regard your neighbor's gain as your own gain
and your neighbor's loss as your own loss.*

T'ai Shang Kan Ying P'ien

Are there identifiable ingredients of a successful training program? Certain elements that must be present in order for the program to succeed?

That depends on how you judge success. It also depends on your level of expectation. What you put into your training program will determine what you get out of it. One thing's for certain, if you keep doing what you're doing, you'll keep getting what you get.

So what do you want?

You want to comply with the law? You want to have clean inspection reports and avoid citations and fines? You want to keep your employees from suing you for health problems or the government from suing you for environmental ones? You want to reduce accidents and injuries? You want to avoid the evening news? You want to stay in business?

If a combination of all of these things is what you want, then what you need, whether you realize it or not, is for your workplace *to improve in a variety of areas.* Your training program will not meet these expectations unless improvements are made, unless things change for the better at your company—from management-employee communication to storage and handling of raw materials and wastes.

What does it take to have a successful training program? It takes the same things required to run a successful company or head a successful household. These ingredients are not complicated or technical. They are mysterious only if they have been forgotten. They are difficult to find only because the simplest answers are so easily overlooked.

CONSISTENCY

In any endeavor or any relationship, consistency is crucial.

If you want to improve your racquetball game so that you can compete in the club tournament, you have to practice your backswing consistently, three nights a week for the next six months. If you skip a Monday night here and a Wednesday night there and try to double up on Saturday, not only will your improvement be slowed, but you will probably end up injuring yourself. *Consistent practice is the only way you will improve.*

If you want to provide a stable environment for your children, you have to be consistent in your demands of them. You can't allow them to deviate from established rules one day because it's more convenient or because you're too tired to correct them and then, several days later, punish them for the same deviation—at least not unless you want a revolt. *Consistent behavior requirements are the only way to elicit consistent behavior.*

The warning signal of a troubled relationship: "You're so inconsistent! Today you say one thing, yesterday you said something else. Tell me what you want…"

What employees say when they've caught on to you: "He's your best buddy when no one's around. But as soon as *his* boss shows up, forget it. It's as if he never knew your name…"

Sound familiar? Inconsistent behavior leaves a trail of suspicion in its wake. Because, quite simply, inconsistency is dishonest. And everybody can tell. Sooner or later the people affected by it begin to suspect you of hidden agendas and ulterior motives. Unable to predict how you will respond to their behavior, they will cease to trust you, begin to resent you, and may even come to fear you. Since they cannot trust you to stand by your word, how can they predict when you will turn against them?

When we are inconsistent with ourselves, we risk failure, disappointment, and injury. But we hurt only ourselves. When we are inconsistent with others, not only do we jeopardize our relationship and whatever endeavor we are engaged in together, but we risk hurting that person in the most stubbornly slow-healing method possible: *by betraying their trust.*

If this happens to your training program, only time, and lots of it, can heal the wounds.

Consistency is essential to the training process for the same reasons it is essential to any other human interaction. And that's what your program must be inextricably rooted in: human relationships:

Between employees and their jobs.
Between employees and the materials and equipment they handle.
Among coworkers.
Between employees and their supervisors.
Between supervisors and ultimate decision makers.

> Training begins with the condition of these relationships *right now.*
> Training is evaluated on the basis of how these relationships *improve.*

How can they improve if the framework for the relationship itself, the *training program,* is inconsistent? Your own experience holds the answer: *they can't.*

In what ways does your program need to show consistency? Following are some areas.

Content

Consistency in content means that all the parts of your training program must adhere together in harmony. You must strive to eliminate contradictions in the procedures taught through the training program. For example (as you will see in Chapters 7 and 8), Hazard Communication training must be consistent with Hazardous Waste training, and both must be consistent with on-the-job training for anyone handling hazardous materials or wastes. Each part of the program must be supported and maintained by the other parts, not only of the program, but of the procedures for the operation of the company.

In order to achieve this kind of coherence and congruity in your program, you can't take a piecemeal approach. Whether one person or several are involved in managing the program and conducting training, it is *essential* that you step back and look at the big picture. You must see your company as one interlocking unit, where inconsistency is disastrous and cannot be tolerated. First, determine all the different types of training that need to be conducted. Then assess the training that is already being conducted. Examine all training content to make sure it isn't contradictory. Evaluate the content to make sure it is aimed toward achieving a common goal.

Procedures

Consistency in procedures requires two things to be true. First, it requires that your procedures are non-contradictory, that safety procedures taught in a training class don't contradict standard operating procedures taught on the job. Second, it requires that your procedures, once established, are followed on an on-going basis, day-in, day-out, with as little deviation as possible.

In order for procedures to be followed consistently, all managers must know about them, stand behind them, and *follow them.* Even if you alone are "in charge" of the safety and environmental training program at your company, you can't do this part alone. Unless following the procedures established to meet safety and environmental compliance requirements are integrated into the other managers' responsibilities, they won't be followed consistently. And you will be virtually powerless to change that. You may think that this would only be a problem at large companies, but the truth of the matter is that the size of the company is irrelevant.

Case in Point: At a small print shop with 10 employees the foreman has been given responsibility for safety and environmental compliance and training. He sets the procedures and makes sure they are followed *except* when the owner comes back to run a special job. When this happens (about 2 or 3 times a month) the owner, who knows nothing about his compliance obligations (he hates that stuff; that's why he gave the job to his foreman), operates the equipment, heedless of the foreman's procedures. What kind of message does this send to the pressmen who watch "the boss" work carelessly while they must operate under strict procedures regarding the use of respirators, goggles, and machine guards or else risk disciplinary action?

Support

Consistent support of a training program means several things, because your program needs support from several areas.

It needs support from you, the person "in charge" of the program or conducting the training, in the sense of *follow-through*. This means that, if you say you are going to do something that relates to training, whether that means answer a question, obtain a different kind of protective equipment, or see that a certain piece of equipment is repaired, you must do it.

It also needs support from the rest of the company management, especially upper management, in the sense of *commitment*. This means that they will provide the resources that will enable you to build a consistent program.

Finally, it needs support in the sense of continual verbal and visual *reinforcement*. Issues relevant to safety and environmental compliance must become part of the everyday language of communication at the company. Safety and environmental concerns must be talked about all the time—so often, in fact, that working *without* taking these concerns into consideration becomes unimaginable, part of the past. It must be virtually impossible to get away from the safety and environmental goals of the company, because they exist everywhere: in meeting topics, bulletins, memos, posters, announcements, theme days, incentive programs, and so on.

In order for your safety and environmental training program to improve the workplace, it can't be confined to meetings or classes. It can't be a responsibility foisted onto one safety manager or environmental compliance manager and conveniently ignored by everyone else. And it certainly can't be addressed only at compliance deadlines and then shoved onto the back burner the rest of the year. The training program will not work unless it is supported by everyone at the company, all of the time.

Enforcement

Consistency in enforcement means that the same procedures apply to *everyone, all of the time*. You can't look the other way when it's a long-term employee or your best friend doing something wrong. Neither can you let things slide just because it's a busy time of year or everyone's been working overtime and is tired.

Do you remember being a kid and getting your room ready for a visit from a grandparent or some other distinguished personage? Suddenly your mother would be after you like a drill sergeant, telling you where to put things that normally had no place to be put. It probably made no sense to you—all this flurry, all these changes for one person, one visit. Maybe you resented the intrusion into your status quo. Maybe you just shrugged your shoulders and rolled your eyes, submitting to events you were powerless to control. Either way, it's doubtful your heart was in it—whatever your mother *made you do.*

Or do you remember thinking your parents or teachers had a "pet," and it wasn't you? It was an older or younger sibling or the cutest, smartest, largest or smallest kid in the class who got away with murder. If they were late coming home, everyone was concerned. If you were late, everyone was mad. If they cracked a joke in class, the teacher was amused. If you tried to be funny, the teacher sent you to the principal's office. It seemed that a different code of conduct existed for them than existed for you. It never did make any sense to you. And you *never liked it.*

Consistent enforcement of rules of conduct are as important now as they were then. When failure to follow safety procedures is winked at or completely ignored throughout most of the year, but becomes a firing offense when an inspector shows up, what does that tell your employees about your attitude toward the procedures or, for that matter, *toward them?* Are you sending them the right message?

COMMUNICATION

"What we have here is failure to communicate." A funny line from the movie Cool Hand Luke, it's not funny at all in the workplace.

What is communication, anyway? It's more than just talking. Communication demands *reciprocity*—a give and take that is key to your training program. In order for communication to occur, you have to *listen* to the employees you're training. And they have to be *unafraid to speak* their minds to you.

Outside the workplace, communication is a lost art. People are either too busy to talk or too knowledgeable to listen. The nonthreatening exchange of information and ideas is all but unheard of. It seems that everyone believes they have too much to lose to risk exposing what they do or don't know. In the workplace, this is a lot to overcome, especially when it's compounded by barriers of authority and, sometimes, race. But it must be overcome. The future of your company depends upon it.

Why is communication so important? Because it is the only vehicle you have for determining where you are, where you've been, and where you're going with respect to your safety and environmental training. Unless you have open lines of communication, not only between yourself and your students, but also between yourself and the ultimate decision makers at your company (the folks with the purse strings), you will have no control over the program.

This is, of course, the case in all kinds of human interactions—parents and children, husbands and wives, employers and employees. When communication is stifled by ego or fear, the relationship is stunted and knowledge suffers.

In order to have your training program avoid stagnation, it must be continually evaluated—not only from the perspective of how much it costs and how much it saves, but from the *inside,* from the point of view of the employees receiving the training. What they tell you is the best information you have for judging whether or not your training is meeting their needs.

But it is unlikely that they will speak honestly unless they are confident that in giving input, making suggestions, sharing ideas, criticizing, assessing, and evaluating the effectiveness of the training they are risking no harm to themselves and will experience no retaliation for their comments. They must feel that negative comments will not cost them their job. There must be a strong trust between you and your students that frees them to be honest. A training program will not work if everyone is afraid to tell the emperor he has no clothes.

And once the emperor's condition is brought to general attention, action must be taken to change it. True communication always *results in action.* Unless you and the other designers or managers of the training program are capable of responding to the ideas, suggestions, or needs expressed by the students, their input soon will end.

GOALS

The purpose of communication is to get results, to effect change. But who determines what those results should be? Who sets the goals to be achieved through safety and environmental training?

Ideally, every person at your company is involved in the goal setting process. Goals can exist at multiple levels—for the company as a whole, for departments, or for individuals.

At each level, goals can be as simple or as complicated as safety and environmental needs demand. For instance, a company goal might be to reduce air pollution emissions by a certain percentage; a department goal might be to have perfect inspection reports for hazardous materials container management; an individual goal might be to feel better by wearing a respirator.

By opening up the lines of communication at the company and assessing needs expressed by personnel at all levels, goals can be established that give focus and purpose to training program activities. Goals stimulate interest in the training program by personalizing it and by providing a means for judging its effectiveness.

Goals also aid in team-building because they pull employees together for a common purpose. This is where incentives can play a valuable role in your training program. The purpose of incentives is *not to give people a reason for adhering to the policies and procedures* established by the training program. *Incentives are not bribes.* The purpose of incentives is to give employees a system for evaluating their progress toward goals that *they have had a hand in establishing.*

Often, however, incentives are *imposed upon* a flagging safety program like sugary frosting smeared on a stale cake. Rather than communicating with employees to find out why the program isn't working and to set goals and incentives to make it work, managers come in with a fancy song and dance, promising everything from silk jackets to jam boxes if employees will kindly jump through the following predetermined hoops. Incentive programs must begin with goals; and goals must begin with *the employee being trained.* If your program has no goals, it is stale, and no amount of expensive bribery will make it fresh again. At least not for long.

Any human endeavor without goals is a lost endeavor. A sense of futility creeps in, as people can't help but ask: What's the *point?* When no answer to this question is forthcoming, the endeavor itself is on shaky ground indeed. All of us need mile markers. Remember travelling with your parents and asking, every 15 minutes or so, "Where are we going?" and "How much longer 'til we're there?" How much different is that from sitting down with your managers at the end of the year and setting production goals for the first quarter of the following one? Have you ever sat down with your bank book and your budget and figured out how, if you save x-number of dollars over the next few months, you can go on vacation, donate to a charity, or buy a new sofa? The goal becomes *the point of the activity,* whether that activity is travelling, operating a business, or going to work every day. The goal provides not only a reason for the activity, but it lends a vitality, an *essentialness,* to the activity itself.

When reaching goals is the point of your safety and environmental training program, each person at your company will learn that it is in their best interest to subordinate personal prominence or ego to the common good. Eventually, everyone, from the company president to the janitor, will recognize their equal membership in an interdependent unit: the company. Each member has a role that is integral to the operation of the whole. With this membership comes an equal responsibility, not only to enforce safety and environmental policies and procedures (as in the case of the company management), or to follow safety and environmental policies and procedures (as in the case of employees), but to constantly review those policies and procedures for utility and relevance.

Why? Because the perspective of each member is unique. What is obvious to a production worker regarding a safety issue is never observed by a business manager. In order for goals to be relevant and mile markers recognizable, they must be developed with each perspective in mind.

DISCIPLINE

Discipline is a frightening word to most people. It implies a rigid adherence to rules and codes of behavior. It conjures a tenacity that is sober and serious, even painful. Definitely not fun.

Having discipline means having self-control. Having the ability to commit to something, to stick to it, and to behave in a consistent manner in order to achieve or obtain it, even if doing so is uncomfortable or difficult. Discipline manifests itself in the physical world; it is rooted in behavior. You are either disciplined or you're not; the proof is not in what you say, but in what you do.

Behaving in a disciplined manner is easier if there is some purpose to it, if you have some goal in mind. The goal is the motivation for the commitment, and thus the discipline that must follow to reach the goal.

Perhaps you want to run in the New York Marathon. That's your goal. If you are serious about achieving that goal, then you will commit to it by adhering to a training regimen that includes running, swimming, weightlifting, and controlling your diet. You will gauge your progress by timing your runs and by routine monitoring of your heart rate and weight.

Commitment is a mental activity. Discipline is its physical manifestation. Discipline is what transforms commitment into achievement.

It is easy to accept that runners, ballet dancers, and football players must have discipline in order to succeed in their respective fields, because they are seen solely in a physical context. But what about you, what about other managers at your company, and what about the students in your training classes? What sort of discipline do you need?

It goes without saying that you must have a company-wide *commitment* to your safety and environmental training program in order for it to work. You've heard that before, right? In order for the program to get off the ground, all the managers and owners have to pledge their support of it. But this commitment actually occurs on an intellectual, almost hypothetical level, doesn't it? Your commitment and theirs is only as powerful as wishful thinking unless you have the discipline to *act on it*. Discipline is required to make training work.

Where does it come from? Discipline must start at the top, at the highest level of company management. It can manifest itself in a variety of ways, all of which have a direct influence on whether or not the program will meet established goals.

1. Management can show discipline in allocating funds for training services, protective equipment, and other workplace improvements related to safety and health. When it might be easier to say there is no money available, management can turn commitment into reality by *making money available,* even if it requires going without or tightening belts in other areas.
2. Management can show discipline in the enforcement of safe procedures. Managers need to control instincts that tell them production should take precedence

over everything else in the workplace. Even if short-term production speed is jeopardized, managers must enforce the adherence to safe procedures. Regardless of the extenuating circumstances of deadlines, backlogged orders, or the experience or skill of a given employee, managers must be disciplined in demanding safe work practices from *every employee, all of the time*—no shortcuts and no exceptions. Remember, your employees will only be as disciplined as you are.

As discussed in Chapter 4, in the context of protective equipment use, lack of discipline (or perhaps more correctly stated, *fear* of discipline) is the primary reason that training programs fail. Managers complain that they have "tried everything" to get their people to wear respirators or hearing protection and "nothing works." How about a little disciplinary action? OSHA standards provide employers a legal framework within which disciplinary action is very much allowed. If you provide your employees with protective equipment that fits them, *and* you train them in when, where, and why the equipment should be worn, *and* you tell them you expect them to wear it, *and* that you feel it is important that they wear it, *and* you are going to conduct "surprise inspections" to check up on them, *and* that, beginning at some agreed upon date, if you find them not wearing the equipment more than x number of times they will be sent home for three days without pay and after that they will be let go, you are completely within your rights as an employer. You have certain standards by which you must operate your company to protect safety and health. If, given information and fair warning, an employee cannot work according to those standards, it is your obligation (to him or her and to the rest of your employees) to let that employee go.

Are you thinking: "But some of my best workers are my worst safety offenders?" Are you thinking: "My plant supervisor has terrible safety habits, but if I let him go, *then* where would I be?"

Where would you be if he or she lost an eye, sued you for hearing loss or, maimed a hand in a piece of machinery? If someone is your "worst safety offender," then he or she is no longer your "best worker," not by today's standards. Why? Because that person is risking too much.

Change is hard for everyone. Chances are that, in order for your training program to effect real change in your workplace, some people in your organization are going to have to suffer a rude awakening. You may not like it any more than they do. It may embarrass you as much or more than it does them. But that's just too bad. Because in order for your training program to become more than empty exercise, behavior must be disciplined.

Think of it this way. Wouldn't you prefer your favorite line supervisor to suffer a rude awakening, even if it means losing his or her job, than suffer an accident or injury and lose his or her health?

3. Management can show discipline in following safe procedures. As mentioned

in the section on consistency, managers often have one set of standards by which they judge employee behavior and another by which they judge their own. They may demand that employees wear safety glasses at all times while they wear only prescription lenses. They may "write up" an employee they observe working without a respirator and then do the very same job without ever putting one on. Managers can accomplish a tremendous amount toward the success of the training program simply by following the procedures they are supposed to enforce.

All the best laid plans and policy statements, the most well-intentioned written programs, the most energetic training classes ultimately will accomplish very little at your company unless accompanied by not only commitment but the discipline to see it through. It takes a lot of stick-to-itiveness to implement a training program. It is not easy to prioritize and emphasize, to compel and urge, to fail, fall down, and get back up again. But it is the only way to realize your goals, and *theirs*.

LIFE

It is a dead household—a household without challenge, without change. It is a place where fear reigns and where self-interest rules every action. Have you ever been in a home like that? Or a classroom? How about a relationship? Or a company?

It feels as if there is *no life in it.*

It is as if the people in such situations have died emotionally and intellectually. Sure, their bodies are still there, going through the motions. But they have become like turtles, expending as little energy as possible, ducking back inside their shells at the slightest hint of risk. In such situations, people are not living; they're trapped.

Life is in flux. It is, by definition, not a stagnant proposition. Why then should we expect the institutions of life to be permanent and inflexible? But we do. We don't like things to change, from our relationship with our kids to the way we've always done things at work. We'll fight life tooth and nail trying to contain it, control it, clamp down on it, and keep it the same. And when we do that, we always end up hurting ourselves. Because life always wins.

So, when it comes to your safety and environmental training program, why not try *letting life in on the front end?* If you can predict life will win in the long run, it makes a lot of sense to get it on your team from the beginning. In practical terms, what in the world does this mean?

It means accepting that the program must grow and change.

A stagnant program is a dead program. In order to be useful to the company, the program must be alive. It must have as its goals the satisfaction of wants and needs. It must be realistic and relevant to the moment. It must be malleable and expandable, correctable and amendable. If the training program doesn't grow as the company grows, then the company suffers. And the program, of course, is worse than worthless.

For example, nearly all safety and environmental training compliance requires the development of written programs. These programs are supposed to represent the basis upon which training is conducted. They are supposed to be *living documents*. Procedural guidelines that are routinely reviewed, updated, and evaluated to insure that they reflect what is actually going on in the workplace. The existence of the program in written form actually is supposed to encourage this kind of creative analysis of the status quo to see how and where it might be improved.

Often, however, the opposite result is achieved. Written solely for the purpose of getting something on paper to comply with a regulation, the program dies even as the words hit the page. Once encased in a notebook, collecting dust on a shelf, it becomes a kind of shrine to compliance, never to be reviewed again. The first and last creative thought regarding the regulation it was written for died with it the day it was placed on the credenza behind your desk.

The unthinking inflexibility represented by this kind of approach to written programs only serves to encourage the belief that safety and environmental training is irrelevant and pointless, that it is based on arbitrary requirements that have nothing to do with reality, that it is, in effect, dead—all of which can be made true if it is all you expect. But if you expect training to deal with reality, then, like life itself, your training program must be vital and dynamic. It must encourage creative expression without fear of punishment. It must welcome change. It must move steadily toward a higher level of quality and wellness.

LEADERSHIP

Somebody has to be in charge, bottom line. When the program fails, somebody has to be the fall guy. When it succeeds, somebody has to stand up and explain what it took to bring it this far.

The leader who is capable of building a successful safety and environmental training program can be almost anyone. Stereotypes don't apply. This person can be young or old, male or female, highly educated or a high school graduate, funny or quite serious, an excellent speaker or somewhat of a drone. None of these things matter. Only two specific traits are required: *patience* and *passion*.

First, *passion*. What is it? It's feeling, emotion, zeal. It's raw energy fueled by belief, rooted sometimes in nothing more tangible than gut instinct. Passion transcends the intellectual. It overwhelms the logical and sensible. It ignores the easy or predictable.

When faced with can't, passion sees *can*. When told no, passion hears *maybe*. When the world presents obstacles, passion embraces *opportunities*. When an employee says "I won't," passion thinks: "Another motivational challenge!" When management says "You've lost your mind," passion responds: "Things are finally beginning to improve around here."

So what does this have to do with your training program? *Everything*. Unless you believe, *with a passion,* in the value of training to individual employees, to the company, and to the community, you will find it difficult, if not impossible, to withstand the slings and arrows that are guaranteed to come your way. If not from nay-saying management, then from nay-saying employees. From one group or the other and sometimes from both at once, you are bound to hear "can't," "won't," "no," and "you've lost your mind" enough times to make even the most ardent optimist weak in the knees.

That's the time when only passion will keep you going. Because anyone who did not *believe beyond reason* in the essentialness of the program would simply take a hint and throw in the towel. But as the leader of the program, such capitulation is not allowed.

Patience is the other side of passion. It separates leaders from zealots, revolutionaries from rabble rousers.

In order for the training program to succeed in meeting the goals you and the rest of the company have established for it, it must endure. And it can't endure if it's leader is *burned out,* or if he or she has been alienated from those that he or she is supposed to help because they *haven't changed fast enough.* The program can survive without a leader, true. But a leader who is disappointed, angry, and resentful is worse than no leader at all. Passion can do this to you if you're not careful.

How do you avoid it?

Put all your sincerity, all your heartfelt belief, all your energy, and all your hopes and plans for what can be accomplished through the training program on one side of a scale. *And then fill up the other side with patience.* To carry the program through the long haul, this much, plus just a little more, is how much patience a leader needs.

3

The Right-to-Know

Working nine to five, what a way to make a living
Barely getting by
It's all taking and no giving
They just use your mind and they never give you credit
It's enough to drive you crazy if you let it

"9 to 5"
Dolly Parton

Most of us recognize the Right-to-Know Law as a training regulation. It is that, true. But it is also a guideline to better management. How? In order to comply with the requirements of the Right-to-Know Law, you have to *believe* your employees have a basic, inalienable right to knowledge of the safety and environmental hazards of the workplace. Lip service, posturing, and pretending won't work. You have to *believe*, at a gut level, that openness builds trust, that real communication with your employees makes you stronger not more vulnerable. You have to *believe* that the more your employees know, the more ownership they will have in their work, and the more of themselves they will invest. You have to want that investment, not because you *think* you should want it, but because you *know* it is essential.

In order to comply with the Right-to-Know Law, you have to *believe* that the strongest tie that binds you and your employees is not what you know that they don't, but what you trust your employees enough to tell them. Not so surprisingly, this belief is central to effective management in any context.

As you study this law and implement its requirements, notice its influence. You may find it sneaks up on you, creeping into every nook and cranny of your role as manager, persistently rattling bones in closets filled with assumptions about what it means to be a manager and an employee. In time, you may discard those old attitudes and replace them with a new model built around the Right-to-Know.

For every training regulation it can be said that it is the employer's responsibility to provide information and the employee's right to learn, own, and utilize it. The bottom line is this: If you incorporate the Right-to-Know attitude into your management style and accept the shared responsibility and commitment that is essential to successful training, then your entire training program and every regulatory compliance effort will mean something at your company. In time, company culture itself will be revolutionized.

POWERFUL POSSIBILITIES

The Right-to-Know attitude originates with the Right-to-Know Law or OSHA Hazard Communication Standard, found in Title 29 Code of Federal Regulations Part 1910.1200. This law simply states that every employee who works with hazardous materials has a *right to know* about the hazards of those materials.

Traditionally, the role of the Right-to-Know has been underestimated for two reasons: because it has been viewed solely in the limited context of a law to be complied with, and because the rewards of compliance have been seen as limited to protection from "enemies" of business and industry—those who will "get you" if you don't comply.

This perception has been revealed as destructive and shortsighted because it deprives employers of the power, available through an open, trusting, proactive management style, to use regulations to make tangible improvements in the workplace.

And the Right-to-Know is designed to do just that. As soon as you decide to equate compliance with progress, the Right-to-Know becomes a creative manager's powerhouse—a workshop full of tools you can apply to your workplace as needed. When you begin thinking this way, it becomes clear that compliance with the Right-to-Know is not only integral to the effective operation of your business; it is essential as the basis of every other subject in your safety and environmental training program.

Why does the Right-to-Know hold so much power? Three reasons:

1. It gives your employees a *right*.

    What is a right, anyway? We hear the word a lot—my rights, your rights, right-of-way, civil rights, right to privacy, right to life.

    A right is a power or privilege that belongs to someone, usually by law, culture, or tradition. Where there are rights, there is struggle and controversy. Why? Because as soon as one group claims a right, there is another group ready to misinterpret, resent, fight over, ignore, or abuse it. Clearly, rights are serious business.

    A law that gives employees a right to knowledge about hazardous materials would seem to empower the employee—and it does—but it also empowers you, the employer or trainer, with the role of champion of the right. When you stand before 1 employee or 100 and tell them they have a right to know about the chemical hazards of the workplace, you've taken up the gauntlet as their champion, because the news came out of your mouth. It's toward you that they will look for that right to be fulfilled.

Like most leadership roles, this one has its advantages and disadvantages.

On the upside: you're in control of the information, how it's presented, and how it's applied. If you have particular safety or health concerns related to hazardous materials, such as respirator usage or storage of flammable liquids, you can use Right-to-Know as a more effective venue for getting your message across by presenting it in the context of a *right to be exercised* rather than as the topic of just one more safety meeting. As mentioned earlier, with knowledge comes ownership and investment. The culture of blame can be replaced with the culture of shared responsibility, even shared vision, as individual rights are championed, exercised, and reinforced.

On the downside: if you tell your employees they have a right to know but you don't take it seriously, you make a half-baked attempt to provide information or you don't follow through, they may call your bluff. If you can't fulfill their right, you're in trouble. Because, after all, the right to knowledge is guaranteed to them by law. So, for your sake and theirs, before you begin telling employees they have the Right-to-Know, you need to expect to be called upon to fulfill it, and you need to give some serious thought to how you're going to fulfill it. What are you going to say? What information, resources or equipment are you prepared to provide? How quickly and consistently are you prepared to follow through? Responding to your employees needs under the Right-to-Know provides opportunity to build trust or destroy it. How important is it to build a culture of openness and communication between you and your employees?

As a manager or trainer, you are empowered to champion a right; employees are empowered to exercise a right, *to know.* Knowing is not easy either. After all, there are so many kinds of knowledge:

a. To know the facts is to have a clear understanding of the facts.
b. To know the way to Chicago it to be sure of how to get there from here.
c. To know the alphabet is to have the alphabet securely in your memory.
d. To know how to repair an air conditioner or operate a printing press is to possess a skill as a result of both study and experience.
e. To know right from wrong is to be able to distinguish between them.
f. To know how a chemical will behave in your body is to be able to apply general information to a specific situation.

To know anything is not simple. It doesn't happen overnight. Gaining knowledge is like buying an extremely valuable item on layaway. You put a little down this month, a little next month, and a little the next, until finally your efforts culminate in ownership. To own something is to possess it as personal property.

When you really know something, however, your ownership is much more permanent than that of a house, a car, or an expensive coat. All of these things can easily be taken away or destroyed. However, when you know some fact or circumstance, such as your name, the smell of sulfuric acid, or the feel of acetone on your skin, it is an ownership not easily taken away.

The Right-to-Know empowers employees with the opportunity to own information about the hazardous materials they handle on the job. Why opportunity? Because, as with any right, it's up to the person who has the right to exercise it. Knowledge cannot be legislated, *only the opportunity to obtain it.*

The Right-to-Know requires the employer to provide the employee specific tools to help him or her learn about the materials he or she works with: labeling systems, Material Safety Data Sheets, a written program, training. It's up to the employer to present information that is both accurate and valuable, and to provide a work environment that is conducive to learning and applying new information. It's up to the employee to learn and apply the information provided, staking his or her claim as owner of it.

2. It connects work to life.

You may be thinking, no, Right-to-Know is about hazardous materials in the workplace, period. If you are, then you're seeing the trees, but missing the forest. You're also missing an opportunity to further empower your employees and yourself.

It's true, Right-to-Know provides a framework for teaching employees about the hazardous materials in their workplace. In other words, the law applies only to those materials your employees are exposed to at work, not to every hazardous material in the world. But if you limit information regarding these materials only to what occurs at work, within the confines of a 40-hour week and four walls, you disassociate the material from the employee's life (and your own).

For most people, the time they spend "at work" will never be as important or meaningful as the time they spend "at life." Therefore, if you want them to learn something about the hazardous materials at work, you'd best connect whatever you tell them with the quality of their life outside work. Quality of life encompasses not only physical health, but environmental health as well.

The Right-to-Know law doesn't come right out and say this, of course. But it's implied in the law's training content requirement and in its performance-oriented directive.

The law says that, as a result of training, your employees are supposed to be able to find, understand, and apply information about how the hazardous materials at their workplace or in their workarea will behave in a variety of situations. Depending on the job and level of interaction with a hazardous material, employees might need to know: how it will behave in contact with their skin, when inhaled, or ingested; how it might impact their health over time; how it will behave in contact with other materials; how it will behave under different storage and handling conditions; if it will burn, explode, or ignite other materials; what to do if it spills; and how to dispose of it safely.

Although it is certainly possible to address all of this information as sterile textbook data, as the facts, figures, and statistics of a job unconnected to the interesting nitty-gritty of life experiences, doing so is rather like milking a bear for a glass of milk. It takes one heck of a lot of effort. And it's not worth much once you get it.

For instance, you can tell an employee that acetone is toxic and flammable. You can explain the Threshold Limit Value, flashpoint, and vapor density of acetone. You can tell him or her to keep lids on containers of acetone and to put out ignition sources when acetone is being used. You can tell the employee that, when hand pumping acetone into a jug, he or she is supposed to wear gloves, goggles, and a respirator for protection from the acute and chronic health effects of skin absorption and inhalation.

Once you've done all of this, you'll be tempted to think you've done your duty. After all, you've told your employee what the law requires. But does that employee *own* the information you've provided? Does what you've said mean enough to *influence his or her behavior?*

How might the learning process be changed if you focused on *why* the employee should change his or her behavior when handling acetone? What if you illustrated each "why" with an example, not from work, but from life? Would your employee listen closer if you related the toxic reactions overexposure to acetone produces to the drunkenness experienced after drinking too many beers? What if you connected chronic health effects from chemical overexposure to cirrhosis of the liver from alcoholism, would his or her ears perk up then? Would he or she be more diligent about cleaning up spills if it was understood that acetone is among the organic solvents his or her spouse is so concerned about keeping out of groundwater? Would he or she put tops on containers more faithfully if it was realized that acetone vapors pollute the air outside the workplace as well as inside? Would he or she be less careless about smoking around solvents like acetone if it was understood that they could ignite as quickly as the gasoline used to light a bonfire?

Part of the power of the Right-to-Know lies in its inherent mandate to break down the work/life dichotomy where hazardous materials are concerned. If you miss this aspect of the law, you've missed its essence.

The message inherent in the Right-to-Know is very personal. It says that the impacts of hazardous materials on our bodies and our environment (inside and outside work) are not random events, but events within the power of every person who handles these materials to *predict* and *prevent.* It follows then that knowledge of chemical behaviors is the link between people's actions and the consequences of those actions, a link that affords them the opportunity to gain some amount of autonomy and control over those consequences, if they choose to exercise their right to do so.

Your employees will spend at least half of their lives at work. In championing their Right-to-Know you might emphasize that the degree to which they choose to exercise it is the degree to which they are, in effect, investing in a kind of life insurance. Insurance that when they retire they will carry with them no negative health effects. Exercising the Right-to-Know is insurance that pays out in quality of life beyond work.

Viewed from this perspective, it is clear that compliance with every regulation impacting a facility, not only OSHA regulations, but those of other agencies as well, would be made easier if every employee impacted by that regulation were told that he or she had a Right-to-Know about it, and if, in the context of the Right-to-Know, the activities he or she was trained to perform in compliance with every safety or environmental regulation (hazardous waste, transportation, air pollution, water discharge and so on), were connected to the employee's life, and the life of his or her family, out in the *real world*.
3. It is need-based.

The Right-to-Know gives employees the opportunity to obtain and own information about the hazardous materials they handle in their job. It also personalizes technical information so that employees can relate it to life beyond work. Further, it mandates that employee training and information are sufficient to enable them to work safely and to protect themselves from chemical hazards.

In other words, training must be based on need.

So, what do employees *need to know?* And which employees need to know it?

The law states that employees who work with hazardous materials need to know about the health and physical hazards of those materials. This statement raises three questions:

1. What is a hazardous material?
2. What is a health hazard and a physical hazard?
3. What kind of knowledge do people who work with these materials need to have about health and physical hazards in order to work safely?

OSHA defines a hazardous material as a material listed in one of a variety of publications, including Appendix A or B of Title 29 CFR Part 1910.1200; Subpart Z of Title 29 CFR Part 1910; Threshold Limit Values for Chemical Substances and Physical Agents in the Work Environment by the American Conference of Governmental Industrial Hygienists; Annual Report on Carcinogens by the National Toxicology Project; or the Monographs of the International Agency for Research on Cancer. (See the Appendix for more hazardous materials resources.)

If you're not interested in memorizing the names of thousands of chemicals, but you are interested in providing your employees with information they can understand and use, then you might think of a hazardous material as a material that, when used, poses measurable long- or short-term health (body) and/or physical (environmental) hazards (risks).

So, what is a health hazard? The Right-to-Know law says that a health hazard is a chemical that enters the human body through absorption, inhalation, or ingestion and has a negative effect on it. These effects can range from dry skin or a mild rash to lung damage, cancer, or death.

A physical hazard has nothing to do with the body. It is a material that is hazardous because of the way it will behave under certain environmental conditions, such as temperature, humidity or moisture, pressure, shock, contact with an ignition source, or contact with other chemicals. (For specific hazard categories and definitions, see the Glossary.)

What kind of knowledge do employees need to have about these hazards?

Do they need to be able to recall the definitions by rote as they would the alphabet or multiplication tables? Do they need to be able to distinguish between them, in the same way they distinguish between right and wrong? Do employees need to be able to recognize chemical hazards by sight, smell, or physical sensation, in the same instinctive way they recognize colors, the odor of foods, or the sensation of heat, cold, or illness? Do they need to have a general knowledge of how to work safely with these chemicals, in the same way they know how to safely drive a car?

Obviously, different employees need to know different information at different levels, depending on their job description and the chemical interaction their job requires. Since the law is need-based, it can be applied on a sliding scale. The employee with the greatest chemical interaction needs the highest level of knowledge. The employee with little or no chemical interaction, who nonetheless works at a location where chemicals are used or stored, needs general awareness rather than specific knowledge.

The Right-to-Know is designed so that all information provided through training is aimed at meeting the need to know of the employee being trained. There is no such thing as a one-size-fits-all Right-to-Know training program. Neither is there a standard program to meet the requirements of any other safety or environmental training regulation. Only when training is aimed at filling a *specific need to know* is it meeting its intended purpose under the law.

This targeted approach enhances the power of training because it all but insures that the training will be effective. Or, put another way, it all but insures that something will happen as a result of the training. Discussion will be stimulated. Thought will be encouraged. Needs will be examined. Changes and improvements will be made. And that's the point. Employers don't waste time developing or buying training programs that don't fit their employees' needs.

Employees don't get turned off to training because everything they hear relates to them.

Case in Point: Meet Harvey. Harvey is a mechanic. His employer tells him he has a Right-to-Know about the hazards of the chemicals he works with. Harvey is unimpressed. Yeah, I have a right to free speech, too, he thinks. But around here that'll get me fired. Then, Harvey's employer proceeds to conduct a training seminar that does two things Harvey has never encountered before. First, the trainer personalizes everything he says about chemical behaviors so that Harvey can relate to it. For the first time, chemistry doesn't seem boring to Harvey. Actually, he finds himself truly interested in why certain chemicals he's been around all his life act as they do. Second, every chemical and work situation the trainer talks about comes straight out of Harvey's work day. The bad habits and short cuts the trainer discusses are Harvey's bad habits and short cuts. The risks the trainer explains are all Harvey's risks.

"It's your privilege under the law to learn as much about these chemicals as you need to protect yourself right now and in the long run," the trainer says. "It's my duty to provide you with information. It's your right to use it. The choice is yours."

He didn't plan to, but Harvey is listening.

THE NEED TO KNOW

The Right-to-Know is powerful because it gives employees a right to information, it establishes that work is indeed part of life, and it is designed to be applied as needed. It encourages the development of a management style that is respectful, trustworthy and trusting, and open to change and improvement.

It is practical and logical for Right-to-Know or Hazard Communication training to serve as the base upon which all other training is conducted because, regardless of the name of the regulation the training complies with (Lockout/Tagout, RCRA Hazardous Waste Management, Personal Protective Equipment, Waste Minimization, or a host of others), the purpose of the training is the same: to establish an employee's Right-to-Know about, and thus protect him or herself from, a particular workplace hazard, to improve the quality of life away from, as well as at work, and to provide information relevant to the employee's situation—information that he or she needs and can use.

The Right-to-Know law says that you have to assess what an employee needs to know about the hazardous materials in his or her workarea before you can begin training. The mandate to use training to meet individual employee safety needs is very clear. And it makes sense, doesn't it? Not only for Right-to-Know or Hazard Communication training, but for every other training regulation. Too often, however, this is not what happens. Training is conducted to comply with the law, not to meet employee needs.

When you adopt the Right-to-Know management style toward training, the first step in every training compliance effort becomes assessing what your employees need to know.

The Job Description

The job description is the trainer's best friend. Why? It's simple. If you're trying to figure out what any employee needs to know about any workplace hazard, you'll save yourself a lot of false starts and frustrations if you begin with a clear understanding of what he or she does all day.

Small companies are notoriously bad about this, although no company is immune. Everybody knows what everybody does, right? Who has time for descriptions when there's work to be done?

Well, you need to make time. A complete, accurate, written description of each of your employee's jobs is not only essential to your training program, it's an invaluable tool for assessing the efficiency of your organization, and if done right, it can be a morale booster.

Job descriptions are too often confused with job titles. A title is a label: Press Operator, Machinist, Welder, Foreman, Shipping Clerk, Secretary. The label enables us to place the employee in a general job category. The label might or might not let us know what kinds of materials the employee works with. One thing is for sure: Any information provided by the job title is going to be general at best.

By contrast, a job description is a detailed account of what an employee does to earn his or her paycheck. There is no such thing as a standard job description because no two physical facilities are identical and no two employees at different facilities (even if they have the same job title) are managed alike.

The best way to get accurate, complete job descriptions is to go into the facility and ask people what they do. At first they might look at you like you've slipped a cog. "Gosh, Jerry, we've worked together for ten years. I thought you'd have caught on by now." Just laugh and tell them you want to hear it in their words so you can get it on record.

Have them tell you what they do on an average day, from when they arrive until they go home. Also ask them if there are any other tasks they do only occasionally, such as during equipment maintenance or cleaning, for special product orders, or if they have to fill in for someone in another area. Write it all down. Even the grumpiest employee will get a boost knowing that someone took the time to find out what he or she does.

Once you have accurate job descriptions, you can assess the need to know for each employee by identifying the level of interaction he or she has with a particular hazard.

For Right-to-Know or Hazard Communication training, you would assess the employee's interaction with hazardous materials or chemicals. For Lockout/Tagout training, you would assess the employee's interaction with hazardous energy. For protective equipment or respirator training, you would assess the employee's interaction with materials that require the use of protective gear. For RCRA Hazardous Waste training, you would assess the employee's interaction with hazardous waste. For Emergency training, you would assess the employee's potential for interaction with various plant emergencies, such as fires and spills.

As an example, let's look at the four main categories of chemical interaction that most job descriptions fall into:

1. Working in a building where chemicals are present.
2. Handling closed chemical containers.
3. Transferring chemicals between containers.
4. Using chemicals in a hands-on process.

Let's take an employee in each of the four categories and determine what he or she needs to know about hazardous materials.

Secretary at a Manufacturing Company

Sue works in the front office at a furniture manufacturer. Most days, she has no reason to go back into the plant. This is fine with her because she has no idea what's back there and she's afraid of all of it. Occasionally, however, she has to make a trek to deliver a message to someone who won't answer a page. Also, she's been known to go to the lunchroom out in the plant to get Cokes when the machine in the front office is out of order.

Sue's need to know is at the *awareness level.*

Right now Sue knows nothing about the materials or operations in the plant. To her, the terms "chemical" or "hazardous material" are in themselves frightening. Since she knows nothing about degrees of hazard, to her all chemicals are worst-case deadly. Sue's fear makes her a danger to herself and others every time she walks through a production area.

By contrast, if she had an awareness of the types and uses of chemicals at the company, Sue would be more confident and less fearful, thereby posing less risk to the company as a whole.

What does Sue need to know? First, she needs to be told that hazardous materials are used at the company. She needs to be told, in a general sense, why the materials are called hazardous. If most of the materials are flammable, like gasoline or paint thinner, she needs to know it. If most are corrosive, like lye or battery acid, she needs to know it. If she's afraid that hazardous means radioactive or deadly or she has some other misconception, her fears need to be allayed. It will help her if the solvents and corrosives in the plant are compared to those she buys herself and uses at home.

Sue also needs to be able to recognize areas of the plant where chemicals are stored. For instance, she should know that acids are stored in totes, which are the square containers on legs at the south side of the building. She should know that flammable liquids are stored in 55-gallon drums, which are the barrel-shaped containers in the storage room outside the wood finishing department.

Awareness of hazardous materials requires a general understanding of terms such as liquid, solid, gas, vapor, hazardous, corrosive, and flammable. It also requires an ability to associate chemicals, in a general sense, with certain containers or storage areas within the plant.

Warehouse Worker at Plating Company

Joe works on the shipping dock at an electroplating and mechanical plating company. Most of the time he is running a forklift, unloading unplated parts, like nails, and then loading them again after they're plated. Several times a month, however, he has to handle a load of chemicals. The company uses a variety of liquid and solid caustics and acids, chrome, copper, zinc, brighteners, additives, and a host of other materials purchased only on rare occasions. Joe certainly knows a drum of chemicals from a drum of nails. From listening to the platers talk, he has a peripheral understanding that some of what he handles is "bad stuff," whereas some of it is "just like soap." But he couldn't really say which is which.

Joe's need to know is at the *recognition level*.

Right now, since Joe cannot link a chemical container to the specific hazard of the chemical in the container, he is likely to handle all containers the same. This lack of information puts him at risk in the event that a container leaks or if its contents come in contact with heat, flame, incompatible materials, or his own body. Sure, in a given situation Joe might use common sense and guess the correct course of action. However, his lack of information leaves him unprepared to make the informed decisions that would improve his odds of reacting safely.

What does Joe need to know? First, when he is unloading a shipment of drums, he needs to be able to link the chemicals to their hazards on the basis of words, symbols, or even container color or shape. For instance, he needs to be able to recognize the symbol for corrosive and know where containers with that symbol should and should not be stored. If a drum of corrosive material is leaking, he should know to avoid stepping in it or getting it on his skin. Sometimes information needs to be even more specific. For example, if the company uses nitric acid, Joe needs to be warned that it is highly hazardous. He needs to know that it is stored in stainless steel barrels, and he needs to be able to recognize those containers on sight. If the company handles a water reactive material, he needs to know the symbol that indicates water reactivity. If a container of sodium hydrosulfite, for example, is leaking or smoking, Joe needs to know to stop what he's doing immediately and get help.

Recognition of hazardous materials requires an ability to associate words and symbols with specific chemical behaviors.

Film Processor
Many personnel, including drivers of tanker trucks, chemical storage tank farm operators, and workers at rail terminals, transfer chemicals from container to container. The size of the container is unimportant, as long as the primary interaction with the chemical is limited to transferring it.

Liz is a film processor in the art department of a large commercial printing company. On a routine basis, she has to go to the store room and select the proper chemicals to add to the film processing equipment. She has to place the chemicals in the processor, either by manually pouring them into the proper receptacles or by hooking up an automatic feed system. She has a pretty good grasp of which container to pour where, but she's not as clear on which chemical is more hazardous or why.

Liz's need to know is at the level of *identification and discrimination.*

What does she need to know? Although her physical contact with chemicals generally is minimal, unlike Joe the dockworker, she needs to be able to tell the difference or discriminate between chemicals that may have the same general hazards or symbolic labeling. She needs to be able to connect a specific chemical name with its use. She also needs to have a more detailed understanding of the hazards of specific chemicals than either Sue or Joe. For instance, she needs to be able to distinguish between a flammable and a combustible and between a corrosive and an irritant. She needs to know which chemicals she should keep from contacting one another. Finally, Liz needs to know what to do if a chemical is spilled.

Discrimination requires an ability to identify and distinguish between chemicals based upon their names and uses and to associate hazard warnings, such as flammable, combustible, and corrosive with degree of hazard.

Car Body Painter
Hank is a painter at a large suburban car dealership. His job requires daily, almost constant contact with hazardous materials. From mixing and preparing paints to spraying paints to cleaning paint guns and other tools, he is nearly always elbow deep in a variety of chemicals. Additionally, he works side by side with other people who are welding, sanding fillers, and using adhesives. Hank knows which chemicals are which. He also has a pretty good idea which chemicals are flammable, which will give you a "buzz" if you breathe them too long, and which will irritate your skin. What he isn't as clear on is *why* they'll do these things. Because he's so familiar with the chemicals, Hank hasn't given much thought to whether or not his particular method of using the chemical is putting him at risk of being hurt or becoming ill. And he certainly hasn't spent a lot of time pondering how his interactions with chemicals today might impact his future.

Hank's need to know is at the level of *application.*

What does he need to know? Like Liz the film processor, Hank must know one chemical from another. But more important than being able to recognize a flammable liquid from a canister of acetylene or a corrosive wire wheel cleaner from a combustible degreaser, Hank must be able to apply his knowledge of a chemical's behavior to his own interaction with and exposure to the chemical. This requires both a higher level of understanding of specific chemical hazards than is needed by the secretary, dockworker, or film processor, as well as an ability to relate abstract, technical information to the realities of daily life.

For example, Hank needs to know that the lacquer thinner he uses as a gun flush is flammable. But, in order to protect himself on the job, his knowledge can't end there. Hank also needs to know that the lacquer thinner evaporates very quickly when exposed to air, creating vapors that are heavier than air. These vapors will gather on the floor and will move about, seeking a source of ignition. Since the flashpoint of thinner is about 50°F, Hank needs to know that a spark, a heater, or even a cigarette can ignite the vapors, even when they are distant from the drum or can. He needs to understand that when the vapors ignite they will travel back to their source, most likely exploding the container. Applying this information to his daily work, Hank will know to minimize the amount of time that lacquer thinner is exposed to air. He'll know to turn off equipment and engines when he is using thinner and to check around the shop to be sure that no one else is smoking or welding.

In the same way, Hank must be able to connect information about the toxicity of a paint or coating with his own daily exposure to it. He needs to be able to recognize when he is overexposed, identify bodily sensations that indicate overexposure, and actively take steps to avoid or minimize those sensations by changing work habits or wearing protective equipment.

Assessing the need to know is the hardest part of compliance with any training regulation. Figure 3-1 illustrates four levels of need. Placing your employees into one of these four levels of chemical interaction is just one method of assessing what they need to know to work safely. It doesn't matter how you determine what they need to know, as long as you make an accurate determination and design your training program to fulfill it.

Designing a training program without first considering exactly what your employees need to know about workplace hazards is rather like sitting down to write a term paper before you've picked a topic or conducted any research.

Figuring out your topic, conducting the research, and outlining your approach may be the hard part. But once you've done it, writing the paper is a breeze.

DECIPHERING THE REGULATORY CODE

What does the Hazard Communication Standard Require?

The Right-to-Know Law gave birth to the Hazard Communication Standard, which is Part 1910.1200 in Title 29 Code of Federal Regulations. Unless your state

The Right-to-Know   41

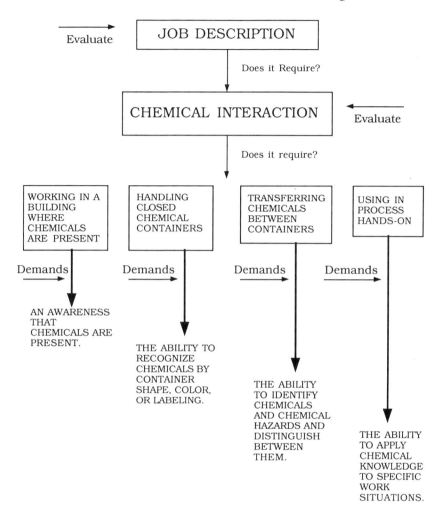

Figure 3-1.   Need-To-Know Flowchart.

has added to the Federal Regulation, you should look to 1910.1200 to find out exactly how the regulation requires you to provide your employees with information regarding the hazards of the chemicals in their workplace.[1]

---

[1] Several states that have obtained jurisdiction over OSHA programs, have adopted 1910.1200, but made certain additions or enhancements. States cannot subtract from the Federal Regulation. If your company operates in one of these states, contact your State Department of Labor for a copy of your state Right-to-Know law.

For purposes of compliance with 1910.1200, an "employer" is a person, including a contractor or subcontractor, engaged in a business where chemicals are used, distributed, or produced for use or distribution. The law allows no exemptions based on the number of employees an employer has. If you have one employee who may be exposed to hazardous materials under normal operating conditions or in foreseeable emergencies, you are covered by this standard.

The Hazard Communication Standard specifies four tools that the employer must use to get chemical hazard information across to employees: The Written Hazard Communication Program, Labels and Other Forms of Warning, Material Safety Data Sheets, and Employee Information and Training. As we go through these requirements, many may seem impractical or even laughable if taken at face value. As you read you may think: "Ha! That will never work at my plant. Whoever wrote this regulation ought to get out more often." And that's fine.

But one word of advice: The secret to understanding this regulation and *any other regulation* with training requirements is recognizing that the requirements are *not intended to be taken at face value*. Quite the contrary, it is intended that you will apply them to your particular workplace in *whatever form will get the job done,—that is, in whatever manner works best for you. The requirements are designed to be molded into a system that makes sense. But you* have to do the molding.

Declaring that this regulation, or any other, is a bunch of useless nonsense and then putting together a program that is, in fact, a bunch of useless nonsense doesn't let you off the hook. The whole purpose of training regulations is to help you in actually improving your employees' safety. In order to do this you have to get personally involved. There's no way around it.

This book establishes the Right-to-Know as the basis upon which all other training programs should be built, and so the Hazard Communication Standard is used here as a model for analyzing and interpreting regulatory requirements. But it cannot be overemphasized that, for *all regulations-mandated training,* you have to use the requirements as a guideline and adapt them wherever necessary to the needs of your employees.

Tool #1: Written Hazard Communication Program

Regulatory Code:

> 1910.1200(e) Employers shall develop, implement, and maintain in the workplace a written hazard communication program for their workplaces which at least describes how the criteria specified in paragraphs (f), (g) and (h) of this section for labels and other forms of warning, Material Safety Data Sheets, and employee information and training will be met.

English: The employer is instructed to think through exactly how he or she is going to provide accurate chemical hazard information to his or her employees, and then write a plan for meeting those obligations. Basically, the law requires employers to ask themselves several who, what, when, where, how, and why questions:

- Who at your company will be responsible for inspecting hazardous materials to insure that they are properly labeled?
- When will the inspection be conducted?
- How will improper labels be corrected?
- Who will be responsible for obtaining Material Safety Data Sheets?
- If Material Safety Data Sheets don't arrive with initial shipments of hazardous materials, how will they be obtained?
- Where will they be kept?
- When will they be updated?
- Who will conduct employee training at your company?
- What information will be covered?
- Who needs the information?
- Why do they need it?
- What methods or materials will be used to train employees?
- When will training be conducted? How will it be documented?
- Where will records be kept?

The Written Hazard Communication Program must also include several other specific pieces of information.

Regulatory Code:

> 1910.1200(e)(i) and (ii) A list of the hazardous chemicals known to be present using an identity that is referenced on the Material Safety Data Sheet (the list may be compiled for the workplace as a whole or for individual workareas); and the methods the employer will use to inform employees of the hazards of nonroutine tasks and the hazards associated with any chemicals contained in unlabeled pipes in their workareas.

English: The chemical list is fairly self-explanatory. It bears mentioning, however, that the chemical list should be compiled first, as a hazardous materials inventory. In other words, you can't very well know if you have Material Safety Data Sheets for every chemical in your workplace until you've conducted an inventory to find out what you have. Walk around your workplace and write down the name on the label of every chemical product you find. Look in closets, cabinets, underneath tables, in the basement, and in the storage shed. Then go to your Material Safety Data Sheets and make sure you have a sheet for every chemical on your list. If you follow this procedure, you're much more likely to end up with an accurate list than if you build your list from the Material Safety Data Sheets you currently have on hand.

The instructions regarding nonroutine tasks and unlabeled pipes are actually training content reminders. Nonroutine tasks are unusual activities that may come within the realm of an employee's job description, but that he or she doesn't do on a regular basis. Nonroutine tasks might include the cleaning of reactor vessels or tanks and maintenance or cleaning of equipment or machines. Unlabeled pipes could contain almost anything, from natural gas to lubricating oil to a wide range of process chemicals.

These two particular sources of workplace hazards have been identified as being significant enough to warrant their own section of the Written Program. In other words, just listing Nonroutine Tasks and Unlabeled Pipes as part of your training class content isn't enough. As the employer, you must again ask yourself pointed questions:

- What are nonroutine tasks at this company?
- What are the hazards of those tasks?
- Who is required to undertake them?
- What procedure will we follow to teach our employees about those hazards?
- Will we cover the information as part of our annual training, or will we hold job-specific safety meetings before these tasks are begun?
- Do we have any unlabeled pipes?
- What chemicals are in those pipes?
- What are the hazards of those chemicals?
- Who works in areas where unlabeled pipes are present?
- Is there any way employees could be exposed to those hazards?
- Can the pipes be labeled?
- How will we cover this information in our training classes?

Tool #2: Labels and Other Forms of Warning

Regulatory Code:

> 1910.1200(f)(1) The chemical manufacturer, importer, or distributor shall ensure that each container of hazardous chemicals leaving the workplace is labeled, tagged or marked with the following information: (i) identity of the hazardous chemical(s); (ii) appropriate hazard warnings; and (iii) name and address of the chemical manufacturer, importer, or other responsible party.
>
> 1910.1200(f)(4) requires that if the chemical is also regulated by OSHA under a substance-specific health standard, the chemical manufacturer, importer, distributor or employer must make sure that his labels meet any requirements of that standard.
>
> 1910.1200(f)(5) Except as provided in paragraphs (f)(6) and (f)(7), the employer shall ensure that each container of hazardous chemicals in the workplace is labeled, tagged or marked with the following information: (i) identity of the hazardous chemical(s); (ii) appropriate hazard warnings.

1910.1200(f)(6) The employer may use signs, placards, process sheets, batch tickets, operating procedures or other such written materials instead of affixing labels to individual stationary process containers, as long as the alternative method identifies the container(s) to which it applies (and provides the identity of the material(s) in the container(s) and the appropriate hazard warnings). The written materials must be readily available to the employees in their workarea throughout each work shift.

1910.1200(f)(7) The employer is NOT required to label portable containers into which hazardous chemicals are transferred from labeled containers, AND which are intended only for the immediate use of the employee who performs the transfer.

1910.1200(f)(8) The employer shall not remove or deface existing labels on incoming containers of hazardous chemicals, unless the container is immediately marked with the required information.

1910.1200(f)(9) The employer shall ensure that labels or other forms of warning are legible, in English, and prominently displayed on the container or readily available in the workarea throughout each work shift. Employers having employees who speak other languages may add the information in their language to the material presented, as long as the information is presented in English as well.

1910.1200(f)(10) The chemical manufacturer, importer, distributor or employer need not affix new labels to comply with this section if existing labels already convey the required information.

English: 1910.1200(f)(1), (5), (6), and (7) are the crux of the matter where hazardous materials labeling is concerned. These paragraphs place the responsibility for insuring proper chemical container labeling on the employer. The purpose of labeling, according to this Standard, is to provide employees with pertinent hazard information regarding the contents of chemical containers in the workplace. However, since the Standard is performance-oriented, you can choose whatever labeling method or method of information transfer that works best at your site—whatever is most practical.

The term "appropriate hazard warnings" sounds awfully vague. The intention here again, however, is to allow employers to provide performance-oriented labeling, or information that is useful to (and can be actively used by) their personnel. In the Standard, "Hazard Warning" is defined as "any words, pictures, symbols or combination thereof appearing on a label or other appropriate form of warning which convey the hazard(s) of the chemical(s) in the container(s)." "Appropriate," in this case, should be read as "whatever fits the situation and makes sense."

There are two principles at work here:

1. *Labels must provide information*—First and foremost, a label is a communication tool. If the words, pictures, and symbols on a label are unintelligible to an employee, then they are not providing information. They are not "appropriate." Far too often, however, employers use labeling systems that their employees either cannot read or cannot apply. These employees learn to view a label not

as a tool for their protection, but as a colorful, meaningless, decoration "required by law."

Case in Point: Businesses of all kinds use Hazardous Materials Identification System (HMIS) labels. In theory, this is a great system. Each container of hazardous materials is labeled with a sticker with three color bars. The blue bar is for HEALTH, the red bar for FLAMMABILITY, and the yellow bar for REACTIVITY. At the bottom of the label is a white section for PROTECTIVE EQUIPMENT. At the right-hand side of each bar there is a blank box. In order to fill in the boxes, which must be done in order for the label to have any meaning, you have to go to a wall chart that is supposed to be posted in the workarea. The wall chart will tell you how to rank the health, fire, and reactivity hazard of your hazardous material on a scale of 0 to 4. The chart also provides pictures of combinations of personal protective equipment (gloves and goggles; gloves, goggles, and respirator; etc.), with alphabetical designations A through K. In order for the stickers on the containers of hazardous materials to have any value whatsoever, these numbers and letters must be filled in (teaching people what they mean is another matter). Far too often, however, the boxes are left blank. Employees are taught never to leave a container unlabeled, so they diligently label every container. But the labels are meaningless. They serve no purpose whatsoever and are a total waste of time and money.

This is another illustration of the waste caused by the pervasive belief that whatever is required by a regulation probably makes no sense and certainly isn't useful. Applying the labels is a regulatory requirement. Since they are required by law, no one in the company expects the labels to have any value or any practical use. Therefore, no one in the company questions them, asking what they mean or why they're blank.

2. *Labels must fit use and user*—Second, a label will be more successful in providing meaningful information if it can be made to fit the use and the user of the chemical. In other words, the label must address what employees need to know to work safely in terms they can understand. In order to do this, employers may need to take a step beyond the labels available in the marketplace and design their own.

Case in Point: Housekeeping and laundry chemicals usually are labeled by the manufacturer with labels that are crammed with fine, difficult-to-read print. Often, these labels will use a term such as sodium hydroxide instead of the more familiar term lye. They will say "Warning: Corrosive and Reactive," instead of the more easily understood "Warning: Burns Skin on Contact, Don't Mix with Other Chemicals." When employees have poor literacy skills, complicated labels with fine print and lots of words can be overwhelming. These labels may meet the letter of the law, but they aren't meeting the need to know of the employee. They aren't providing information the employee can understand. Instead of a complicated manufacturer's label, housekeepers and laundresses would benefit from simple labels that directly address their use of and exposure to the chemical, such as "Don't Mix"; "Burns Skin"; "Wear Gloves"; and "Don't Breathe Mist."

Portable Containers
The issue of portable container labeling is always a source of confusion and debate. The purpose of labeling requirements in general is to avoid having unlabeled containers of hazardous materials sitting around the workplace. Unlabeled containers of hazardous materials are dangerous because, if an employee doesn't know what something is, he or she might mistake it for something else, might use it improperly, or might mix it with an incompatible substance. That much makes sense. Now, all we have to do is apply this principle to portable containers. The key to correctly interpreting this part of the Standard is found in the phrase "intended only for the immediate use of the employee who performs the transfer."

If an employee pumps a quart of lacquer thinner into a can, takes it to a work station, and is the sole user of the entire quart of thinner, without the can ever leaving his or her possession, then the can doesn't have to be labeled. However, if the employee doesn't use the entire quart of thinner before leaving the work station for lunch, going to another part of the building, or going home at the end of the day, then the can is supposed to be labeled. In this case the most practical label would read "Lacquer Thinner, Flammable." It might also include a statement regarding the company's policy regarding the use of personal protective equipment while working with lacquer thinner, such as "Respirator Required" or "Goggles Required." Figure 3-2 shows several practical labels.

The Standard doesn't directly address what should be done with the portable container once it is "empty." As anyone who has worked with flammable solvents or corrosive liquids knows, containers used to hold these materials are not empty until they have been triple rinsed. So, going back to the employee who kept the unlabeled can of thinner at his or her side until it was empty, what should be done with the "empty" can to insure that no one will decide to use it as an ashtray or put corrosive liquids in it?

This is where site-specific standard operating procedures would be a big help. Can the employee put the can in a locker or some other location where no one else can get to it? Or can it be put in a community storage location designated and labeled for flammable liquids? Would it be easier just to set a policy that all portable containers must be labeled, thereby avoiding confusion and misinterpretation?

As an employer, how you handle this matter is up to you—as long as you handle it and document your procedures in your Written Program.

48   Safety and Environmental Training

Figure 3-2.   Label Examples.

Tool #3: Material Safety Data Sheets

Most of 1910.1200(g) deals with the requirements for developing Material Safety Data Sheets. The Standard requires chemical manufacturers and importers to obtain or develop a Material Safety Data Sheet for every chemical they produce or import. And it specifies that the Material Safety Data Sheet must be in English and that certain information must be present, including:

1. The identity used on the label;
2. The chemical and common name(s) (if it is a single substance);
3. The chemical and common name(s) that contribute to the known hazards of the material (if it is a mixture);
4. The chemical and common name(s) of all ingredients that are health hazards and that are present in 1% or greater (except for carcinogens, which must be listed if concentrations are at least 0.1%);
5. The chemical and common name(s) of all ingredients that are health hazards and that make up less than 1% (0.1% for carcinogens) of the total mixture, if there is evidence that the ingredient(s) could be released from the mixture in concentrations that would exceed an established OSHA Permissible Exposure

Limit (PEL) or American Conference of Governmental Industrial Hygienists (ACGIH) Threshold Limit Valve (TLV), or could present a health hazard to employees;
6. The chemical and common name(s) of all ingredients that have been determined to present a physical hazard when present in the mixture;
7. Physical and chemical characteristics of the chemical, such as vapor pressure, flashpoint, and so on;
8. The physical hazards of the chemical, including potential for fire, explosion, and reactivity;
9. The health hazards of the chemical, including signs and symptoms of exposure or any medical conditions that are generally recognized as being aggravated by exposure;
10. The primary route(s) of entry into the human body;
11. The OSHA PEL, ACGIH TLV, and any other exposure limit used or recommended by the chemical manufacturer, importer, or employer preparing the Material Safety Data Sheet;
12. Whether the chemical is listed in the National Toxicology Program Annual Report on Carcinogens or has been found to be a potential carcinogen in the International Agency for Research on Cancer Monographs, or by OSHA;
13. Any generally applicable precautions for safe handling and use that are known to the chemical manufacturer, importer, or employer preparing the Material Safety Data Sheet, including appropriate hygienic practices, protective measures during repair and maintenance of contaminated equipment, and procedures for clean-up of spills and leaks;
14. Any generally applicable control measures that are known to the chemical manufacturer, importer, or employer preparing the Material Safety Data Sheet, such as appropriate engineering controls, work practices, or personal protective equipment;
15. Emergency and first aid procedures;
16. The date of preparation of the Material Safety Data Sheet or the last change to it; and
17. The name, address, and telephone number of the chemical manufacturer, importer, employer, or other responsible party preparing or distributing the Material Safety Data Sheet, who can provide additional information on the chemical and appropriate emergency procedures, if necessary.

It is the responsibility of the chemical manufacturer or importer to provide distributors and employers an appropriate Material Safety Data Sheet with their initial shipment, and with the first shipment after a Material Safety Data Sheet is updated.

Obviously, Material Safety Data Sheets are complex, detailed documents requiring a great deal of technical information. The information on the sheet must be accurate to be valuable. However, the value of Material Safety Data Sheets in general is called into question for two reasons:

1. Although OSHA requires the above 17 items to be covered on a Material Safety Data Sheet, they require no specific form for that information. No two Material Safety Data Sheets developed by two different manufacturers are the same. This lack of standard form only compounds the confusion inherent in written technical data when it is presented to generally nontechnical employees.
2. Health hazard information provided on a Material Safety Data Sheet is presented both in a format too esoteric for most people to relate to real life, such as Lethal Concentration Dosages for albino rats or Threshold Limit Values in parts per million, and covered by general, worst-case warning statements that suggest they were added by the manufacturer's legal department. Neither provides much useful guidance for the employee using the chemical.

But the matter of preparing a Material Safety Data Sheet is the subject of another book, not this one. The task at hand is determining what the law requires employers to *do* with Material Safety Data Sheets, in addition to insuring that they are legible, in English, and include the required information.

Regulatory Code:

1910.1200(g)(1) Employers shall have a Material Safety Data Sheet for each hazardous chemical which they use.

1910.1200(g)(8) The employer shall maintain copies of the required Material Safety Data Sheets for each hazardous chemical in the workplace, and shall ensure that they are readily accessible during each work shift to employees when they are in their workarea(s).

1910.1200(g)(9) Where employees must travel between workplaces during a workshift, i.e. their work is carried out at more than one geographical location, the Material Safety Data Sheets may be kept at a central location at the primary workplace facility. In this situation, the employer shall ensure that the employees can immediately obtain the required information in an emergency.

1910.1200(g)(10) Material Safety Data Sheets may be kept in any form, including operating procedures, and may be designed to cover groups of hazardous chemicals in a workarea where it may be more important to address the hazards of a process rather than individual process chemicals. However, the employer shall ensure that in all cases the required information is provided for each hazardous chemical, and is readily available during each work shift to employees when they are in their workarea(s).

1910.1200(g)(11) Material Safety Data Sheets shall also be made readily available, upon request, to designated representatives and to the Assistant Secretary, in accordance with the requirements of Title 29 CFR 1910.20(e). The Director shall also be given access to Material Safety Data Sheets in the same manner.

English: Employers must have a Material Safety Data Sheet for each hazardous material that they use. "Use" is defined by the standard as meaning to package, handle, react, or transfer, which just about covers all industrial and commercial uses of chemicals.

Exemptions
There is no exemption based on the size of the container. Employers at medical and dental offices and laboratories often get hung up on this, thinking that, just because the amount of a given chemical that they store is small, a Material Safety Data Sheet doesn't have to be obtained. OSHA allows no exemption based on amount. Laboratory chemicals definitely are covered by this requirement, as are operations such as warehousing or retail sales, where containers of hazardous materials are sealed when they arrive and are supposed to remain sealed.

The only materials for which the employer does *not* have to obtain Material Safety Data Sheets are hazardous wastes, tobacco, wood and wood products, food, drugs, or alcoholic beverages in a retail establishment that are packaged for sale to consumers; foods, drugs, or cosmetics intended for personal consumption by employees while in the workplace; hazardous materials bought at retail establishments and used only occasionally; manufactured items that do not release or result in exposure to hazardous materials; and drugs intended for direct administration to the patient (tablets, pills).

Usefulness
It warrants restating that employers are supposed to have Material Safety Data Sheets for all of the chemicals they *use*. This does not mean that you should have Material Safety Data Sheets for all of the products your vendors carry or your parent company manufactures.

Often, chemical manufacturers or distributors have books of Material Safety Data Sheets for every product they make or carry. They will give these books to their customers, and their customers will put these books on their shelves. Car dealerships and body shops are especially bad about this. The manager will keep the car manufacturer's parts books on a shelf in his office, thinking he is complying with the law. This is not the case.

Why? To answer that, you have to consider why Material Safety Data Sheets are required in the first place.

The Standard requires employers to have Material Safety Data Sheets for each hazardous material in the workplace so that health and physical hazard data will be available for use by employees during regular work activities and emergencies. If you have Material Safety Data Sheets for every product your vendors distribute, it is unreasonable to think that anyone will be able to sort through all of that on demand and come up with the right sheet for one your chemicals. When you think about it for a minute, it is simply impractical.

Plus, since the Standard requires Material Safety Data Sheets to be accessible during each shift to employees "when they are in their workareas," it only makes sense to cull out the sheets for products you don't have, discard them, and duplicate only the sheets for products you do have for placement in the workplace.

The truth of the matter is, however, that most employers who rely on their vendors' or parent companies' Material Safety Data Sheet books never duplicate all or any of them. They just keep them on a shelf in their office. They don't put together workarea books of Material Safety Data Sheets that reflect what is really present in the workplace; neither do their chemical lists (required as part of the Written Hazard Communication Program) reflect the hazardous materials that are actually present in the workplace. In other words, they don't meet the requirements of the Standard.

The requirement for Material Safety Data Sheets, like every other requirement of the Hazard Communication Standard, demands employer involvement, thought, and input. It requires that you provide not just any old information to collect dust on a shelf but vital information your employees *need and can use.*

Paragraph (10) of this section illustrates this principle better than any other. There is no standard form for the Material Safety Data Sheet. As mentioned earlier, this is a drawback of the Standard, as anyone who has tried to teach their employees how to read and understand Material Safety Data Sheets developed by a variety of manufacturers knows very well.

However, the absence of a standard form allows employers the latitude to *reformat* any Material Safety Data Sheet they receive to make it more appropriate to the workplace and less confusing to their employees. As long as the required information is provided, Material Safety Data Sheets can be streamlined or combined to cover groups of hazardous materials, processes involving hazardous materials, or even specific operating procedures.

Few realize it, but any employer, if he or she wants to take the time and energy to do it, can transform Material Safety Data Sheets from an ineffectual burden that everyone at the company either ignores or fears into a valuable informational and training tool everyone relies upon. The Standard not only allows but encourages you to do this.

Group Material Safety Data Sheets

The Group Material Safety Data Sheet is an excellent tool for classroom training. As an alternative to sifting through a stack of 50 or more sheets, most of which are in fine print, a class of car painters will need to look at only four Group Material Safety Data Sheets—one for Thinners and Strippers, one for Fillers, one for Adhesives, and one for Paints (see Figure 3-3).

Keep in mind that these are not intended to be replacements for manufacturer's Material Safety Data Sheets. The individual Material Safety Data Sheets still must be available, and the painters need to know where they are, but the painters are unlikely ever to use them—basically because they are too hard to read and there is

Group Material Safety Data Sheet

Section I

Product Group: Automotive Paint and Coatings

Names on Labels: Sherwin-Williams Lacquer
                     Dupont Cronar Clears
                     3M Undercoating

Figure 3-3. Excerpt from a Model Group Material Safety Data Sheet.

too much information on them that they don't care about. Spending a lot of time in a training class talking to people about something they are unlikely to use doesn't make a lot of sense.

In contrast, the Group Material Safety Data Sheet provides hazard information that the painters need to know, or are more interested in knowing, in a format they can read and understand. A training class built around a Group Material Safety Data Sheet is more likely to capture their attention.

Group Material Safety Data Sheets for these car painters are more effective at their work stations as safety reminders than is a binder of manufacturer's Material Safety Data Sheets stuffed in a drawer or hanging on the wall. The Group Material Safety Data Sheets can be printed in large dark type and posted on lockers, tool racks, bulletin boards, and in the lunch room. The sheets can be posted individually and rotated around the shop, or all at once at some central location. Or the sheets can be posted in alternation with other safety reminders.

The primary advantage of a Group Material Safety Data Sheet is that it distills all the information on a manufacturer's Material Safety Data Sheet down to its most essential elements—that is, those of greatest importance to the employees using it. To use the same esoteric, remote, and difficult-to-grasp language often found on a manufacturer's Material Safety Data Sheet on your Group Material Safety Data Sheet would be to miss the point entirely. In developing a Group Material Safety Data Sheet, your number one concern should be whether or not it will be of value to the *least educated* employee you have. As much as possible, you should use language *that employee will understand.* Cater to him or her (see Figure 3-4).

Also, in developing a Group Material Safety Data Sheet, make your suggestions and procedures as specific to your workplace and to the work habits of your employees as possible. For example, the painters that this Group Material Safety Data Sheet was developed for have been known to use lacquer thinner to wash their

54   Safety and Environmental Training

Group Material Safety Data Sheet

Section II

Primary Hazardous Ingredients: ORGANIC SOLVENTS

Section III

Health Hazard Information

How does it enter your body?
   Breathing vapors
   Absorption through skin

If you are exposed to too much, how do you feel?
   "High" or "drunk"
   Dizzy or nauseated
   Skin is dry, irritated

What protective equipment should you wear?
   Organic vapor air-purifying respirator
   Nitrile gloves

Figure 3-4.   Excerpt from a Model Group Material Safety Data Sheet.

hands and arms after a job. Additionally, one painter in the shop was found never to have cleaned his respirator and to have used the same cartridges for an entire year. The exhalation flaps actually melted inside the respirator. The Group Material Safety Data Sheet reminds painters to clean and check their respirators once a week and to use the soap provided in the restroom to remove paint from their skin.

There probably isn't a painter alive who hasn't joked about getting "high" on paint fumes. Most also know what it feels like to be "hung over" the next morning. Instead of using remote, technical language, such as Threshold Limit Values, to describe their exposure to paint fumes, this Group Material Safety Data Sheet uses descriptive terms that they can identify with. The Group Material Safety Data Sheet reminds them that every time they feel one of those sensations—high, drunk, dizzy—they are breathing harmful levels of chemicals. Feeling one of those sensations should be an indicator to them that something in the work environment is wrong. Either their respirator is malfunctioning, is the wrong type, or is dirty, the ventilation system isn't working, or some other factor is involved.

Using words on a Group Material Safety Data Sheet that mean something to these painters validates their own work experiences, encouraging them to trust their bodies and instincts.

As with the section on Health Hazards, this last section of the Group Material Safety Data Sheet for automotive paints is extremely simple. It provides no numerical values for flashpoint, vapor pressure, or density. Instead, it gives the painters practical information about how they can expect paint to behave. It will evaporate. Vapors will settle on the ground. Vapors can be ignited by heat, spark, or flame, even if they are distant from the source of the vapor (see Figure 3-5).

It bears restating here that the purpose of the Group Material Safety Data Sheet is not to prove how many fancy words you can use to describe the behavior of paints. The purpose is to present relevant chemical data, not in chemist's terminology, but in words that would make sense to the average sixth grader. If some of

---

Group Material Safety Data Sheet

Section IV

Physical Hazard Information

Primary Warning: Paints are FLAMMABLE.

The vapors from paints are heavier than air and will settle on the ground. They can ignite if they come in contact with heat, spark, or flame from a welding torch, match, or cigarette. The heat, spark, or flame can be across the room from the paint and still cause the paint vapors to ignite.

DO NOT SMOKE in Paint Shop, Body Shop, or Paint Kitchen.

Section V

Storage and Handling

Keep lids on paints at all times. Every minute paint is left exposed to the air it is evaporating. This reduces the quality of the paint and causes a fire hazard. Never store paint next to a heat source or in the sun. When you finish using a can of paint, return it to the yellow Flammable Storage Cabinet.

Figure 3-5. Excerpt from a Model Group Safety Data Sheet.

your personnel are more sophisticated than others and demand more precise data, you always have the manufacturer's Material Safety Data Sheet to turn to. (For definitions of terms found on most Material Safety Data Sheets, see the Glossary.)

Another thing that this Group Material Safety Data Sheet does is to incorporate standard operating procedures, or the way that the painters are supposed to do their jobs, into chemical safety procedures. "Keep lids on paints at all times." "Put paints back in the yellow cabinet when you are finished with them." Handling paints safely is presented here as merely part of doing their job. It is not impractical to follow safe procedures; quite the contrary, it's economical. For instance, keeping lids on containers of paint not only reduces fire hazard in the shop, it also makes the paint last longer. A Group Material Safety Data Sheet written with standard operating procedures in mind can help to break the production/safety dichotomy that allows people to believe that if it's safe it must be bad for production.

Tool #4: Employee Information and Training

Regulatory Code:

> 1910.1200(h) Employers shall provide employees with information and training on hazardous chemicals in their workarea at the time of their initial assignment, and whenever a new hazard is introduced into their workarea. (1) Employees shall be informed of: (i) The requirements of this section; (ii) Any operations in their workarea where hazardous chemicals are present; and (iii) The location and availability of the Written Hazard Communication Program, including the required list(s) of hazardous chemicals, and Material Safety Data Sheets required by this section. (2) Employee training shall include at least: (i) Methods and observations that may be used to detect the presence or release of a hazardous chemical in the workarea (such as monitoring conducted by the employer, continuous monitoring devices, visual appearance or odor of hazardous chemicals when being released, etc.); (ii) The physical and health hazards of the chemicals in the workarea; (iii) The measures employees can take to protect themselves from these hazards, including specific procedures the employer has implemented to protect employees from exposure to hazardous chemicals, such as appropriate work practices, emergency procedures, and personal protective equipment to be used; and, (iv) The details of the Hazard Communication Program developed by the employer, including an explanation of the labeling systems and the Material Safety Data Sheets and how the employees can obtain and use the appropriate hazard information.

English: When to train? "At the time of their initial assignment." That means when an employee is hired and before he or she begins work. "And whenever a new hazard is introduced into their workarea." That means whenever an employee must begin working with a new line of chemicals, whenever his or her job description changes, requiring him or her to be exposed to new hazards, or whenever process equipment or procedures change, thereby creating new hazard exposures.

Make It On-going

Obviously, training is tied very closely to standard operating procedures and day-to-day production. In fact, in order for a company to keep up with its training obligations, many people within the management structure must know about these obligations so that they can do their part to make sure that they are being met in a timely fashion. One Safety Manager can't do it alone.

For instance, when a new person is hired, the Personnel or Human Resources Manager needs to be aware of the fact that part of this new employee's orientation is Right-to-Know training. The Personnel Manager needs to make sure that this new hire is scheduled for training with the Safety Manager before being scheduled to begin work. By the same token, when the Production Supervisor approves a new line of process chemicals, he or she needs to have a meeting with the Safety Manager before the new chemicals arrive on the production floor. Together they can review the technical information for the new chemicals and decide the best way to inform the employees who will be using them of their hazards. If the Production Supervisor transfers an employee from one workarea to another, or changes that employee's job duties, he or she needs to inform the Safety Manager of these intentions so that the Safety Manager can decide whether or not additional training is needed. (For more on integrating compliance with production concerns, see Chapter 8.)

Some state laws have expanded the federal requirement for frequency of training to include the statement "training must be conducted at least annually." The purpose of this statement is to emphasize the point that Right-to-Know training is an on-going process. If you aren't conducting it at least annually you can't possibly be keeping up with the changes in chemicals, procedures, equipment, and personnel at your facility. Plus, even if your entire process remains exactly the same for several years, if you and your employees haven't talked about chemical hazards in 12 months or more, most of what you covered last time has been forgotten. Bad habits and short cuts have crept back into your operating procedures, and probably have taken over. Even if your company is in a state that doesn't have this extra provision on the books, it makes good sense to adopt a policy of conducting Right-to-Know training at least annually.

Make It Personal

What is the purpose of training? It is to keep dangerous habits and risky shortcuts at bay by providing employees with enough information about the chemicals they work with that they will respect them, use them properly, and protect themselves from harm. The purpose of training is to talk about what *really goes on* out in the workplace where chemicals are involved and to identify the *specific hazards* associated with those activities. The Standard lists some specific subject areas that are supposed to be covered during training—nuts and bolts topics, such as the requirements of the law, the location of the Written Program and Material Safety

Data Sheets in the workplace, an explanation of the labeling system that is being used, and how to read a Material Safety Data Sheet.

Right-to-Know training programs that stop here cover the mechanics of the law, but they miss its heart. The heart of the law lies in the statements in paragraph (h) that address the employee in the *workarea*.

The Standard says that employees are supposed to be informed of any operations in their workarea where hazardous chemicals are present. Further, training must include the methods and observations that may be used to detect the presence or release of a hazardous chemical in the workarea. Training must address the physical and health hazards of the chemicals in the workarea. This would imply that training should be not just workplace-specific but workarea-specific:

- Training on Material Safety Data Sheets and labeling should address these topics, not in general, but as they relate to the chemicals or processes of a specific workarea.
- Training on physical and health hazards should include not every physical and health hazard identified by OSHA, but the physical and health hazards of the chemicals and processes of a specific workarea, and how an employee can avoid or minimize exposure to them.
- Training on protective measures should address the standard operating procedures and engineering controls of a specific workarea and how they serve to minimize hazard exposure.
- Training on personal protective equipment should address exactly what equipment the company expects each individual employee to wear in a given workarea or during a given work activity and, most importantly, why that employee needs to wear it. (See Chapter 4 for more on training regarding personal protective equipment.)

It is possible to conduct non-worksite- or non-workarea-specific Right-to-Know training. But employers who do this are only complying with half of paragraph (h) of 1910.1200.

THE NETWORK

The requirements of many other OSHA Standards, as well as EPA and DOT regulations, directly intersect with the requirements of the Hazard Communication Standard. You miss a valuable opportunity to reduce your workload and stop reinventing the wheel when you fail to recognize these intersections. You also miss an opportunity to build a training program that integrates multiple aspects of staying safe and healthy into a cohesive whole.

Why separate Right-to-Know or Hazard Communication training from training about welding safety? Or handling compressed gases? Or storing flammable liquids? Or respiratory protection? Or fire safety? Or asbestos? Or lead? Or lab safety? Or cutting and brazing? Or personal protective equipment? Or using dip tanks or spray booths? Or confined space safety? All of these other OSHA standards involve hazardous materials, as defined by 1910.1200.

It is nearly impossible to talk to truck drivers about the DOT requirements for transportation of hazardous materials without talking about the health and physical hazards of those materials and without reviewing Material Safety Data Sheets and identifying important information on labels. Although hazardous waste is exempt from regulation under the Hazard Communication Standard, how can a generator cover his or her Contingency Plan with employees without talking about the health and physical hazards of the waste created from hazardous materials and how they should protect themselves from these hazards when cleaning up spills?

The tendency is to look at each standard as unrelated to any other. In reality, there is no regulatory mandate to deal with them separately and there are no advantages to separating them.

Depending upon an individual employee's job description and chemical interaction, you might find yourself hard pressed to cover his or her need to know under the Hazard Communication Standard without also covering the requirements of one or more corollary standards. If you let it, your training will naturally integrate itself. All you have to do is document it. (More on combining training requirements in Chapter 8.)

Don't panic! This is not to suggest that you develop marathon Hazard Communication training courses that cover everything at once and keep your employees off the production floor for weeks at a stretch. It is to suggest, however, that you think of every compliance effort, safety meeting, and training class at your company that deals with hazardous substances (both virgin and waste materials) as part of your Hazard Communication Program, and that you present it to your employees as such. This approach is one simple way you can extend the impact of the Right-to-Know attitude to every safety meeting or training session at your company.

So, you're having a fire safety class? Advertise it to your personnel as part of Right-to-Know. Your welding gas company is holding a short session on compressed gas safety? Again, introduce the meeting as part of Right-to-Know. A respirator fit test? It's Right-to-Know. Your hazardous waste hauler is holding a session on labeling and manifesting? It's Right-to-Know. At a routine safety meeting you discuss confined space entry procedures? Remind your employees that the meeting is part of your on-going commitment to their Right-to-Know. DOT training? Tell the drivers it's all part of their Right-to-Know as much about the chemicals they handle as they need to be safe.

What are the advantages of doing this?

First, it helps to establish your commitment to the Right-to-Know as a year-round, day-in, day-out commitment, not something you knock out in a couple hours and then forget about until next year. Second, it continually reemphasizes the role of hazardous substances in each employee's life and reestablishes the need for on-going learning about those materials. Thinking about health and physical hazards is not something they can do for a couple of hours each year and then promptly forget about. Third, the more the shared responsibility for the Right-to-Know is talked about, the harder it will be for either the champion or the receiver of the right to ignore it.

Compliance with the Right-to-Know Law or Hazard Communication Standard is a means to a variety of ends—ends limited only by your company's needs and your personal ambition to see them fulfilled.

What would you like to improve at your company? What changes would you like to make? How big do you want your Right-to-Know network to be? Who or what do you want it to include? If you can tailor a Right-to-Know program to fit the specific needs of your personnel, you can also tailor it to fit your own.

This topic is addressed in depth in Chapters 7 and 8, but it bears mentioning here because addressing specific needs, not only of employees but also of management, is a large part of the reason for adopting the Right-to-Know attitude or of complying with any training regulation.

Why are you training?

You say: To fulfill my employees' Right-to-Know under the law.

Why else?

You say: To give my employees the information they need to know to work safely with hazardous materials.

Why else?

You stutter.

Any other reasons?

You say: To improve the management and operation of this company, so it becomes a safer, healthier place to work?

Exactly.

The Impact

The impact of the Right-to-Know attitude toward training compliance is profound because it demands the active, on-going involvement of both employer and employee. It requires a joint commitment to knowing enough about workplace hazards to live and work safely (see Figure 3-6).

When you use the power of this management approach as a tool for progressive change, soon no aspect of your operation, including your other training and compliance, will remain untouched. Your influence will be like that of a large, expertly crafted net. Like a net thrown by an experienced fisherman, if used to best advantage, it will pull in a valuable regulatory catch. But be patient. Crafting a net takes time.

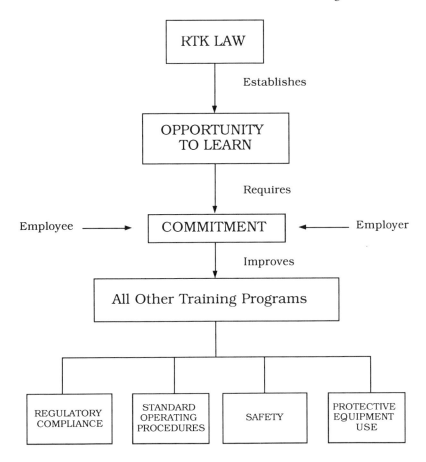

Figure 3-6.   The Impact of the Right-To-Know.

Establishing a Right-to-Know management style that fits you, your employees, and your company won't happen overnight. Leading from trust and compassion and training from respect won't come easy if you've never tried it before. Thinking of yourself as responsible for fulfilling your employees rights may not appeal to you at first. Developing training that is meaningful, that deals with reality, and that meets employees wherever they are with the sole purpose of helping them may seem too awesome a job to tackle all at once.

And that's okay.

What your company and your employees need is not an overnight sensation so much as on-going commitment and discipline. But you have *to let them know* you're committed.

You would laugh at a fisherman who worked hard to make a net and then stood on the beach, never throwing it into the surf. Yet, converting to a Right-to-Know attitude toward training and keeping it to yourself, or applying it only to compliance with the Hazard Communication Standard, or neglecting to discuss your approach with other managers, is just like standing on the beach with an expensive net, wondering why you aren't catching anything.

4

Tools of the Trade

Why not spend some time in determining what is worthwhile for us, and then go after that?

William Ross

A Hazard Communication Program without personal protective equipment is like a tree without roots. Sooner or later it is going to fall over and die, perhaps hurting someone in the process. Is it really this dramatic? Yes. Because protective equipment, along with standard operating procedures and engineering controls (which often incorporate protective equipment), is the number one means an employee has to protect himself from hazardous materials at work. By itself, knowledge of the hazards is not enough.

Of what value is it to the employee to have knowledge of the health hazards of a chemical if he or she has no personal, proactive method for protecting him or herself from those hazards? Not only is it of no value, it's demoralizing. Putting it another way, of what value is it to an employer to provide his or her employees with information on how to work safely and then deprive them of any practical means of doing just that? Not only is it waste of time, it is a foolish liability.

If, as addressed in Chapter 1, the purpose of training is to influence behavior, then explaining the health hazards of blanket and roller washes to a group of pressmen in a Hazard Communication class without giving them any means for protecting themselves from those hazards is not really training them. Basically, it's wasting time. It's also frustrating, because after the class is over, the behavior of the pressmen will remain unchanged.

However, if the same Hazard Communication class was accompanied by the announcement of a policy regarding the use of gloves, goggles, and respirators by pressmen, and if at the class the pressmen were taught how and when to use the equipment and were allowed to select equipment that fit them, then it would be a true training effort. The announced intention of the class would be that, based on the information presented, behavior not only should but *must change*. And the company would adopt a policy to insure that it did.

If Hazard Communication training that doesn't include protective equipment and an equipment use policy makes no sense, then providing protective equipment without a strong informational program and use policy to support it is equally as wasteful, as well as being potentially destructive to both safety and morale.

"We have a whole closet full of respirators, gloves, glasses, boots. You name it. It's in there. Nobody uses any of it." The Safety Manager at this company echoes a sentiment felt by many of his colleagues: *employees don't care.* You can buy all the equipment in the world for them, but they're not going to use it because they simply don't care about protecting themselves.

What this manager faces is a problem of inspiration and motivation. The employees lack knowledge of why wearing protective equipment is beneficial to them. They don't understand the advantages of wearing protective equipment. Additionally, they are not motivated to wear the equipment because their company doesn't require it. The company has no policy regarding the use of protective equipment. It is not enforced.

A Hazard Communication Program accompanied by a protective equipment use policy will get this company's gloves, glasses, and respirators off a closet shelf and into the hands of employees.

In order to effectively influence the way your employees work with hazardous materials, protective equipment really needs to be viewed as *part of* compliance with the Hazard Communication Standard. It needs to be as integral to your compliance program as the requirements listed in 1910.1200.

WHAT DO OSHA'S PERSONAL PROTECTIVE EQUIPMENT STANDARDS REQUIRE?

Like the Hazard Communication Standard, the Protective Equipment Standards, Title 29 Code of Federal Regulations, Parts 1910.132, 1910.133, and 1910.134, offer practical guidelines that are intended to be applied by employers according to their particular workplace needs. If whatever you are doing to comply with the requirements of these Standards at your company is illogical, impractical, or just generally isn't working, then you can consider your compliance unsuccessful. The intent of these regulations is to influence employee behavior so that they work safer and healthier. If that isn't happening at your company, then you need to reevaluate your approach.

General Requirements

Regulatory Code:

> 1910.132 General Requirements (a) Application. Protective equipment, including personal protective equipment for eyes, face, head, and extremities, protective

clothing, respiratory devices, and protective shields and barriers, shall be provided, used, and maintained in a sanitary and reliable condition wherever it is necessary by reason of hazards of processes or environment, chemical hazards, radiological hazards, or mechanical irritants encountered in a manner capable of causing injury or impairment in the function of any part of the body through absorption, inhalation or physical contact. (b) Employee-owned equipment. Where employees provide their own protective equipment, the employer shall be responsible to assure its adequacy, including proper maintenance, and sanitation of such equipment. (c) Design. All personal protective equipment shall be of safe design and construction for the work to be performed.

English: What is protective equipment? It is equipment that will protect the human body from absorbing, inhaling, or otherwise coming into contact with anything that can cause injury or harm. Hard hats, boots, glasses, goggles, face shields, gloves, aprons, tyveks, dust masks, air purifying respirators, supplied air respirators, and splash guards are all protective equipment.

The regulation says that where protective equipment is needed, three things must happen. First, it must be *provided.* Either the employer can provide the equipment, or the employee can provide the equipment, but the employer must insure, as is only logical, that the equipment provided is the right equipment for the job. In other words, the equipment must be of proper design and materials to protect the employee from the designated hazard.

If an employee buys a welding respirator and wears it in a paint booth, then the intent of this requirement is not being met. Likewise, if an employer provides safety glasses for all employees in a production area and the two employees who hand pump solvents wear the glasses, instead of more appropriate chemical splash goggles, then the intent of the requirement is not being met.

Whether the employer or the employee provides the equipment, it is the responsibility of the employer to verify that it is the *right equipment for the job.*

Second, it must be *used.* That's all the guidance the standard provides—one little word sandwiched between two commas. But within it lies the whole crux of the matter. Where absorption, inhalation, or contact hazards exist, protective equipment must be used. Who must use the equipment? Those persons who might be injured or harmed by the absorption, inhalation, or contact.

Is it enough, then, for an employer to provide the correct and proper equipment if that equipment stays in a closet, cabinet, or drawer, unused by the employees who risk injury through absorption, inhalation, or contact with materials or equipment on the production floor? Is it enough for an employer to insist that their employees buy their own equipment, if that equipment hangs on a hook at their work station or collects dust on top of their locker?

Absolutely not.

How does an employer insure that the equipment provided is used? As stated earlier, he or she starts by educating employees about why the equipment is good for them. But information, facts, and figures alone are rarely enough. It must be followed up with a hard and fast policy. And the employer must be willing to *enforce* it.

Many Safety Managers get hung up on the issue of enforcement. As discussed in Chapter 2, it is an unavoidable fact that enforcing the use of protective equipment often puts you at odds with some of your best, most experienced employees; old timers who scoff at glasses and respirators. But the policy must apply to them, too. No exceptions. Hopefully, it won't come to this, but you may have to suspend or even dismiss an employee you would rather not lose. And for their sake and that of the safety and health of every employee at your company, that is exactly what you have to be willing to do.

Third, it must be *maintained*. Protective equipment must be kept in a sanitary and reliable condition. It must be clean and in working order. And it is the responsibility of the employer to insure that it is, whether or not they provided the equipment in the first place.

Glasses can't be cracked or the lenses scratched, or coated with paint or some other material to the point that the employee can't possibly see through them. Gloves can't have holes in them. Protective clothing can't be ripped. Respirators can't be cracked or torn, the outsides can't be coated with dust and the insides coated with grease. Things must be *reasonably* clean.

But my plant is dirty! you cry in alarm. That's just the nature of our process. It can't be helped.

No problem. OSHA knows that equipment gets dirty while people are working. What they want you to do is tidy it up afterwards, and on a *regular basis*. If your plant is particularly dusty or dirty, OSHA doesn't want you to make cleaning requirements so restrictive that your employees can't work; they simply want you to make sure that your employees clean their equipment *after* work.

How you choose to manage the maintenance and cleaning of protective equipment will depend on the type of operation you have. You may decide to have your employees bring their protective equipment to weekly or monthly safety meetings. You may conduct random, unannounced inspections during the year and then review your findings at Hazard Communication or other hazardous materials-oriented training classes. You may pick one employee at random every week or month, depending on the size of your company, inspect his or her equipment, and then hold a private conference to let him or her know how they're doing.

The exact method you use to manage the maintenance and cleaning of your employees' protective equipment doesn't matter, as long as it's practical. *And it works.*

GLASSES AND GOGGLES

Regulatory Code:

> 1910.133 Eye and Face Protection (a) General. (1) Protective eye and face equipment shall be required where there is a reasonable probability of injury that can be prevented by such equipment. In such cases, employers shall make conveniently available a type of protector suitable for the work to be performed, and employees shall use such protectors. No unprotected person shall knowingly be subjected to a hazardous environmental condition. Suitable eye protectors shall be provided where machines or operations present the hazard of flying objects, glare, liquids, injurious radiation, or a combination of these hazards. (2) Protectors shall meet the following minimum requirements: (i) They shall provide adequate protection against the particular hazards for which they are designed. (ii) They shall be reasonably comfortable when worn under the designated conditions. (iii) They shall fit snugly and shall not unduly interfere with the movements of the wearer. (iv) They shall be durable. (v) They shall be capable of being disinfected. (vi) They shall be easily cleanable. (vii) Protectors shall be kept clean and in good repair.

English: When is eye and face equipment required? Whenever it is likely that use of such equipment will prevent injury.

When do employers have to provide this protection? When machines or operations present the hazard of flying objects, glare, liquids, injurious radiation, or a combination of these hazards. When employees are grinding or stamping metal; sawing or sanding wood; handling high-pressure hoses; welding; pumping, pouring, splashing, spraying, misting, or otherwise transferring or handling liquids. Whenever it is possible to imagine that a foreign material could get into an employee's eye as a direct result of their work activity, eye protection should be provided.

The standard requires that employers make such protection *conveniently available* to employees and that, once again, employees *use* them.

The Right Equipment, Where You Need It
It makes sense that employees are more likely to use what is easily accessible. The number two reason, after "It's uncomfortable" or "I look funny," for not wearing protective eye equipment is that getting it and putting it on takes too much valuable production time. Safety Managers can nip this excuse in the bud if you give some serious thought to the specific activities or areas where you would most like to see employees wearing eye protection.

At many companies the place where eye protection is most needed is wherever chemicals are dispensed. For example, hotel housekeepers are at greatest risk of splashing chemicals in their eyes when they fill their hand-held spray bottles at the chemical dispensing rack, which is usually in the laundry. Therefore, chemical splash goggles should always be right there on the rack itself or just adjacent to it, instead of tucked away in the Executive Housekeeper's office or down in the maintenance supply room.

Wherever the eye hazard is greatest, eye protection should be readily available and easy to use. But don't make the mistake of thinking that one kind of eye protector will suffice for every kind of eye hazard, or that if employees in a certain area wear prescription lenses, then additional eye protection is unnecessary.

For example, your company may have a policy that impact resistant safety glasses are required to be worn by everyone who enters a certain production area—workers, management, and visitors. But if certain employees in that area are involved in the handling of chemicals, then the impact resistant safety glasses are not sufficient to protect those particular employees from the hazards of their jobs. For them you must provide, in a convenient location, chemical splash goggles or perhaps a face shield for their use during specified activities. And, of course, you must make sure the equipment actually is used.

Corrective lens wearers tend to be the worst offenders where eye protection is concerned—workers and managers alike. Why? Several simple reasons. Wearing goggles over glasses is uncomfortable, to say the least, and when you take off the goggles, your glasses tend to stick inside and come off, too. From the perspective of most employees, getting special-made safety glasses with prescription lenses hardly seems worth the effort, especially if their employer won't pay for it. After all, if an employee has never had an eye injury, in his or her mind that is simply proof positive that prescription glasses provide a fine barrier to flying particles or chemical splashes. Employees who work outdoors and "can't see" without sunglasses use the same logic.

The bottom line is this: If their employer does not actively enforce the use of protective eyewear, employees who wear prescription glasses (or sunglasses) will rarely if ever use it. It remains, however, the responsibility of the employer to notice when this happens and to establish whatever policies and procedures are necessary to insure that these employees correctly protect their eyes.

The regulation states that no unprotected person shall *knowingly* be subjected to a hazardous condition. The responsibility falls on the employer to be able to recognize hazardous conditions, know how their employees can best be protected from them, and *not allow* their employees to expose themselves to a hazard without that protection.

There is no getting around it: This is not easy.

First, you have to know which equipment fits which hazard. Equipment manufacturers, consultants, regulators, Material Safety Data Sheets, and books can help you figure this part out. (See the Appendix for more information on resources.) Then it's back to the brass tacks of getting your employees to use the equipment, through education, inspections, policies, and disciplinary action. As stated earlier, it is never enough merely to provide the correct equipment and then, when your employees don't use it, to throw your hands in the air and cry, "Oh, well, I did my part." Your part is not over until protective eyewear is worn by every employee who needs protection from a hazard. And so, as you well know, it's never really over.

The Comfort Factor

Section (2) of the regulation excerpted on page 67 lists specific minimum requirements for protective eyewear. These requirements are merely practical: The eyewear must protect against the hazard(s) for which it is used. Chemical splash goggles that do not fit flush against the forehead, temples, and cheekbones will not provide adequate protection against sprayed liquid. Therefore, they should not be used for protection against sprayed liquid. If employees who work in a plant setting that is primarily outdoors are required to wear safety glasses, it makes sense that the glasses provided should be tinted to provide protection from the sun. If clear glasses are provided, that will only serve to encourage the employees to wear regular sunglasses instead.

The eyewear must be reasonably comfortable and must fit snugly. No glasses are what anyone would describe as *comfortable* when you first put them on. At first, even the greatest looking prescription lenses might make you sore behind the ears or bother the bridge of your nose. But they're yours; they're personal. You picked them out because they fit you. Don't forget, there is a reason this stuff is called *personal* protective equipment. It is supposed to fit the person, the individual. The statement that the eyewear must be reasonably comfortable is referring to the general "fit" of the glasses or goggles on the wearer. Are they too small or too large? If the employee turns his or her head quickly, are they likely to fall off? Are they constantly sliding down the employee's nose? Do they dig into the top of the cheeks? Do they have sharp plastic or metal edges that are causing blisters or cuts behind the ears? Providing employees with eyewear that doesn't fit them is clearly self-defeating. The equipment actually can *interfere with rather than enhance* safety. If you enforce the use of such equipment, you are in for a battle. And you can depend on the fact that your employees will come to resent you, as you yourself would resent someone who forced you to wear glasses that didn't fit.

Protective eyewear must be durable, capable of being disinfected, and cleanable. This is pretty straightforward. It needs to be built to last. Cheap eyewear that falls apart the first time it is dropped is likely to be treated carelessly and have to be replaced more often than a quality product that is built to withstand rough treatment. Perhaps this has something to do with the fact that higher-quality eyewear tends to fit employees better. If the glasses or goggles fit well, employees tend to view them as *theirs,* rather than as the company's generic protective equipment. And so they take better care of them.

Eyewear must be kept clean and in good repair. This goes back to the employer's responsibility for maintenance of protective equipment as established in 1910.132. Again, policies and procedures regarding the cleaning, repair, and replacement of protective eyewear are essential to the employees' use of the eyewear. It is the employer's responsibility to insure that eyewear is properly maintained. As stated above, the system you design to accomplish this is entirely up to you, with the single provision that it gets the job done.

Case in Point: The Purchasing Clerk at a chemical manufacturing company is in charge of purchasing safety equipment for plant personnel. This would be fine except for the fact that, first, everything he knows about safety equipment he has learned from salespeople and, second, the primary selection criteria passed down to him from the President is finding whatever equipment is cheapest.

As a consequence, plant personnel, most of whom work outside and handle liquid chemicals, are provided with the most inexpensive untinted safety glasses on the market—glasses intended to protect against flying particles, not liquid sprays and splashes, and certainly not sun glare. But neither the Purchasing Clerk nor the President understands this. Since the glasses are untinted, they provide no relief from the sun. In fact, they seem somehow to make the constant glare worse.

What does this encourage the plant workers to do? Wear sunglasses, of course, which only angers the Purchasing Clerk, who feels the men are being insubordinate. The Clerk routinely tours the plant, threatening the men who are not wearing the safety glasses he provides. Morale plummets as the plant workers go out of their way to do the opposite of whatever the Clerk demands.

Meanwhile, at weekly safety meetings conducted by an outside consultant, plant workers complain about the uncomfortable, unsafe safety glasses that the Purchasing Clerk is trying to force them to wear. Sure they're wearing sunglasses, they say, but only when they're driving the forklift on the white gravel drive or moving drums on the concrete pad. If they didn't wear the sunglasses, they wouldn't be able to see. How safe would that be? As for chemical splash protection, the men who mix chemicals and must wear respirators say they are receiving very little protection from the glasses because they don't fit over the half mask respirators issued to them by the Purchasing Clerk.

The safety consultant, horrified at the situation, does two things. He brings over two full facepiece respirators for the chemical mixers, and he sends a memo to the Plant Manager, the Purchasing Clerk, the Vice President, and the President, in which he recommends that the company provide full facepiece respirators for employees exposed to chemical splashes and also tinted safety glasses with side shields for all plant employees to use during other activities.

One month later, the Purchasing Clerk has "written up" the chemical mixers a total of six times for wearing full facepiece respirators instead of the required safety glasses. The safety consultant has received no response regarding his memo. The word among the plant workers is that the Purchasing Clerk is trying to get the safety consultant fired.

Unfortunately, this nightmarish story is true, and not uncommon. When company management lacks the knowledge or insight both to lead the personnel running the safety program and to listen to the advice of hired experts, even the best laid plans can run amuck, especially when human ego gets involved.

All For One and One For All
At this chemical manufacturing company, the Purchasing Clerk is using personal protective equipment as an opportunity to wield power, to demonstrate his toughness and control over the plant workers. Why is he doing this? It is in the hope that the President will recognize his management prowess and make him manager of the whole safety program. The Purchasing Clerk does not want to be Safety Manager because he is *concerned about the safety of the plant workers.* He wants the job because it would give him a *promotion in power.* The Purchasing Clerk is a shrewd salesman who has managed to wrap the President around his little finger. Since the President is even more ignorant than the Clerk when it comes to safety and protective equipment, he is likely to let the Clerk have his way.

Whenever safety is used as a pawn in the jockeying for power and position, safety is not served. And the personnel whose safety is supposed to be protected will suffer. There are no ifs, ands, or buts about it; they will suffer.

The only way employers or managers can prevent this from happening is to get involved and stay involved in the safety program. Keep everyone connected with the program—the person who buys the equipment, the person who conducts the safety meetings, the person who conducts inspections, and so on—accountable not only to you but to each other. Hold regular meetings. Coordinate decision making. Go out into the plant and ask the hourly worker on the production line if the equipment being used is comfortable, if it works. Get the advice or input of outsiders in selecting equipment. Don't be afraid to change, to try new equipment, or to admit when something just didn't work.

You may be one of the lucky employers whose personnel are committed to the common good, who recognize that what is good for the whole of the company is best for them. If you are, then you don't have to be as watchful. However, if your workplace is more representative of the mainstream, then some of your employees are so caught up in their own glorification and the protection of their own power, however great or small it might be, that they wouldn't recognize the common good if it came up and introduced itself once a day. If you have employees with this type of attitude managing your safety program, then they must have constant direction. To turn them lose with their responsibility is to guarantee that they will transform whatever task they have been given, whether it is buying protective equipment or conducting safety meetings, into grounds for kingship over a realm around which they will strive to build very tall and impenetrable walls. (For more on building a team approach to safety, see Chapter 8.)

RESPIRATORS

Regulatory Code:

> 1910.134 Respiratory Protection (a) Permissible Practice. (1) In the control of those occupational diseases caused by breathing air contaminated with harmful dusts, fogs, fumes, mists, gases, smokes, sprays or vapors, the primary objective shall be to prevent atmospheric contamination. This shall be accomplished as far as feasible by accepted engineering control measures (for example, enclosure or confinement of the operation, general and local ventilation, and substitution of less toxic materials). When effective engineering controls are not feasible, or while they are being instituted, appropriate respirators shall be used pursuant to the following requirements. (2) Respirators shall be provided by the employer when such equipment is necessary to protect the health of the employee. The employer shall provide the respirators which are applicable and suitable for the purpose intended. The employer shall be responsible for the establishment and maintenance of a respiratory protective program which shall include the requirements outlined in paragraph (b) of this section. (3) The employee shall use the provided respiratory protection in accordance with instructions and training received.

English: OSHA makes it clear that the respiratory health of the workplace is a responsibility shared by employers and employees.

Hazard Prevention
According to this standard, the *prevention of contamination* of the breathable air space is to be the employer's first responsibility where respiratory protection is concerned. This is an essential element of the regulation that is often overlooked. But the directive is very clear. The employer's number one duty to his or her employees is *not* to provide them with respirators; it is to do all he or she can to prevent or minimize the air contamination his employees are exposed to.

For example, if you are the owner of a car dealership where, during the winter months, carbon monoxide levels in the service department become uncomfortably high, the regulation says you can't solve that problem simply by requiring your mechanics to wear respirators. You have to install a carbon monoxide collection and exhaust system.

Similarly, if you are the owner of a company that manufactures farm equipment, it is not enough, under this regulation, for you to provide air purifying respirators for your employees to wear as they spray paint equipment parts. First, you must install a paint booth. The idea being that the employee should not have to bear the lion's share of the burden of working in a contaminated air space. That burden should first be the employer's.

Appropriate Protection
However, once all feasible engineering controls are in place, it may be determined, through monitoring and employee symptoms (dizziness, nausea, headache, sinus problems), that respiratory protection is needed. When this is the case, it is the *employer's responsibility* to provide respirators that protect employees from the hazards present. As with any other type of protective equipment, it is up to the employer to figure out which type of respirator will best suit the needs of his or her employees.

If the employees work along a plating line where sulfuric and phosphoric acids and sodium hydroxide are present, they need air purifying respirators with acid gas cartridges. If the employees are welders, they need respirators designed specifically for protection from welding fumes. If the employees work in a carpentry shop, they need dust masks.

Don't be fooled into thinking that one respirator is as good as the next or that any old respirator will do. Each type of respirator is designed to protect against a narrow range of air contaminants. The Material Safety Data Sheet for the material that an employee is exposed to is usually a good place to start in determining what type of respirator is needed. Just remember: A respirator designed to protect against welding fumes will not protect against organic solvents. Neither will a dust/mist mask. And an air purifying respirator that protects against organic solvent vapors will not hold up against acid gases or ammonia.

In addition to providing proper respirators, the employer must also develop standard operating procedures regarding the selection, use, and maintenance of respirators and the education and training of employees. And, once again, it is the *employee's* responsibility to use the respirators provided to them in accordance with standard operating procedures and the training they receive.

Regulatory Code:

> 1910.134(b) Requirements for a minimal acceptable program. (1) Written standard operating procedures governing the selection and use of respirators shall be established. (2) Respirators shall be selected on the basis of hazards to which the worker is exposed. (3) The user shall be instructed and trained in the proper use of respirators and their limitations. (4) Reserved. (5) Respirators shall be regularly cleaned and disinfected. Those used by more than one worker shall be thoroughly cleaned and disinfected after each use. (6) Respirators shall be stored in a convenient, clean and sanitary location. (7) Respirators routinely shall be inspected during cleaning. Worn or deteriorated parts shall be replaced. Respirators for emergency use such as self-contained devices shall be thoroughly inspected at least once a month and after each use. (8) Appropriate surveillance of workarea conditions and degree of employee exposure

or stress shall be maintained. (9) There shall be regular inspection and evaluation to determine the continued effectiveness of the program. (10) Persons should not be assigned to tasks requiring the use of respirators unless it has been determined that they are physically able to perform the work and use the equipment. The local physician shall determine what health and physical conditions are pertinent. The respirator user's medical status should be reviewed periodically (for instance, annually). (11) Approved or accepted respirators shall be used when they are available. The respirator furnished shall provide adequate respiratory protection against the particular hazard for which it is designed in accordance with standards established by competent authorities. The U.S. Department of Interior, Bureau of Mines, and the U.S. Department of Agriculture are recognized as such authorities. Although respirators listed by the U.S. Department of Agriculture continue to be acceptable for protection against specified pesticides, the U.S. Department of the Interior, Bureau of Mines, is the agency now responsible for testing and approving pesticide respirators.

English: Section (1) of this paragraph instructs employers to put in writing a procedure outlining the selection, use, maintenance, and care of respirators at their company. When completed, it should represent *the* definitive statement on respiratory protection at a given company. Because it is so site-specific, there is virtually no *wrong way* to write it, as long as it includes the specific information required by the regulation.

The Written Program

Let's go through the sections that must be included in a written Respiratory Protection Program.

Selection of Respirators
This is easy. The regulation states that proper selection of respirators must be made according to the guidance of American National Standard Practices for Respiratory Protection Z88.2-1969. Often, the recommended respirator for a particular material is listed on the Material Safety Data Sheet. A qualified person at your company or a vendor, consultant, or regulator needs to determine the type of respirator that should be used for each job at your company. (See the Appendix for more information on resources for protective equipment selection.)

The type of respirator specified for each job title or individual employee (by name) needs to be listed in the Program, along with an explanation of how the determination was made and who made it.

Use of Respirators
The amount of information covered in this section of the Program varies widely from company to company, depending upon the type of operation, the severity or variability of the hazards the employees encounter, and whether or not they use respiratory equipment to respond to emergencies.

In general, this section should include the following information:
1. *An explanation of when the respirator should be worn for each job title or individual listed. Sample program excerpt:* "All welders in Department A must wear the respirators issued to them whenever they are welding galvanized steel or whenever they are welding in an area that does not have an exhaust vent immediately overhead."
2. *An explanation of how it is determined that each employee can wear the respirator assigned to him. Sample program excerpt:* "All employees who are issued respirators are required to obtain a physical examination from the company physician, Dr. James Smith, at St. Joseph's Hospital, once each year. The purpose of this exam is to verify that the employee is capable of safely wearing the respirator assigned to him. The Safety Manager reviews the results of the exams with Dr. Smith once each year."
3. *An explanation of how to fit each respirator listed. Sample program excerpt:* "Half-mask air-purifying respirators are fitted by pulling the bottom strap over the head to the back of the neck and pulling the top strap just below the crown of the head. Tighten the straps until the mask is snug over the nose and under the chin by pulling on both ends of the bottom strap and then both ends of the top strap. Check the seal by covering both cartridges with the palms of your hands and inhaling. The respirator should be sucked in against your face. As an alternative, you can also check the seal by covering the exhalation valve in the center of the respirator with your palm and exhaling. You should feel the respirator fill up with air, however, you should not feel any leaks around the edges. If you do, readjust the placement of the respirator and tighten the straps. If after repeated tries you are unable to achieve a seal, contact your supervisor. Never wear a respirator that is not properly sealed."
4. *An explanation of any prohibitions against conditions that prevent a good respirator seal. Sample program excerpt:* "No one who has been issued a respirator is allowed to wear sideburns or a beard that prevent a good face seal. Painters in Department C who have been issued full facepiece respirators cannot wear glasses with temple pieces or contact lenses. Painters who must wear corrective lenses must have glasses mounted inside the facepiece by the Safety Manager."
5. *An explanation of the training and information provided to employees who must wear respirators* (see Figure 4-1). *Sample program excerpt:* "All employees (and their supervisors) who have been issued respirators receive training at the time the respirator initially is issued to them and at least annually thereafter. Training provides an opportunity for each employee to handle his or her respirator, have it fitted properly, test its seal against his or her face, wear it in normal air for a long familiarity period, and wear it in a test atmosphere similar to the one in which he or she will be working. He or she also has an opportunity to practice fitting and adjusting it while being observed by the instructor. At this

```
┌─────────────────────────────────────────────────┐
│  RESPIRATOR  TRAINING  CARD                     │
│  Name: _____  Title: _____  │
│  Height: _____  Weight: _____   │
│  Date of physical: ___ . ___ . ___ . ___ , ___  │
│  Approved for respirator use: ___  ___  ___     │
│       (Safety Manager's initials)               │
│  Respirator assignment: _____         │
│  Date of initial training/Fit test: _____     │
│  Follow-up training: ___  ___  ___  ___         │
└─────────────────────────────────────────────────┘
```

Figure 4-1. Respirator Training Card.

training, employees also are instructed in how to clean their respirator, how to recognize when cartridges need to be replaced, how to inspect it and identify problems, what to do if it malfunctions, and how to store it. Information regarding how a respirator works, the benefits of wearing a respirator, and its role (including its limitations) in protecting health is covered at initial and annual respiratory protection training as well as at all Hazard Communication training conducted at the company."

6. *An explanation of the safe use of respirators in atmospheres immediately dangerous to life and health* (IDLH). Obviously, this section of the program need only be included if there are areas of your operation where, if the respirator failed, the wearer could be overcome by an IDLH or oxygen-deficient atmosphere. *Sample program excerpt*: "Whenever maintenance personnel are required to clean out the vacuum tank in Department F, self-contained breathing apparatus is used. Personnel entering the tank also are equipped with safety harnesses and safety lines. At least two standby personnel are present for each man inside the tank. Standby personnel monitor the safety lines, maintaining contact with the men inside the tank. They have SCBA ready and available for their use in the event that a rescue is necessary."

Maintenance and Care of Respirators

The goal of a maintenance program for respirators should be to insure that your employees are always using equipment that is at its *optimum effectiveness*. The purpose of going to the trouble to write down a maintenance program is to delineate the various components of respirator maintenance and to make specific maintenance assignments—that is, to delegate responsibility.

Like every other part of your written Respiratory Protection Program, your maintenance program must address *your operation*. A manufacturing company with 300 personnel and 15 production departments will need a much different and much more complicated maintenance program than a printing company with 10 employees. There is no standard form, no approach that is wrong, as long as the program achieves the stated goal: Keeping your employees in effective equipment.

There are four basic components of a respirator maintenance program: inspection, cleaning, repair, and storage.

1. Inspection—There are really two types of inspections: *Those conducted by the employee* every time he or she uses a respirator, and *those conducted by the employer* to make sure that routinely used respirators, as well as those kept ready for emergencies, are in good working order.

    Before an employee puts on the respirator, he or she needs to look at it. Is it all in one piece? Is anything ripped or torn? Are valves working? Are cartridges worn out? If something is wrong, it needs to be reported to the supervisor. Every time an employee puts on the respirator, the first thing he or she needs to do is conduct a quick fit test to see if there are any leaks. If he or she cannot get the respirator to fit without leaking, it must be reported to the supervisor. Employees must never wear broken or leaking respirators. They must know that they will not be penalized for taking time away from production to turn in a broken respirator.

    The importance of this last statement cannot be overemphasized. Unless they have been told otherwise, far too often employees will suffer with broken glasses or malfunctioning respirators out of fear that they will be punished or penalized either for breaking the equipment, or for taking time away from their duties to get another one. At companies where the cost of protective equipment is deducted from the employee's paycheck, they sometimes rig the equipment to make it last long past the time it should have been replaced. At one printing company, pressmen use duct tape to hold cartridges in place after the area where the cartridge is supposed to screw in place has cracked. Why? Because they don't want to pay $13 for a new facepiece.

    A strong training and inspection program accompanied, perhaps, by a more equitable system for providing protective equipment, would alleviate this problem.

    Management inspections of routinely used respirators are basically the same as those conducted by employees. Documentation of these inspections is not required by the regulations. However, as mentioned in the discussion of 1910.132, you might find that writing down the results of your inspections and then reviewing them in safety meetings or training classes is an effective method of raising awareness regarding the importance of respirator maintenance. It is also excellent proof that you are doing your job (see Figure 4-2).

    Inspections of emergency equipment are another matter. The regulation requires that respirators and SCBA kept for emergency use are inspected after each use and at least monthly. The dates of the inspections, as well as any findings, must be recorded.

**SMITH PRINTING COMPANY
RESPIRATORY PROTECTION PROGRAM
QUARTERLY INSPECTION**

Date:_____ Inspector's Name:_____ Title:_____

Number of respirators inspected:_____ Department:_____

Examine each respirator and ask yourself these questions. If the answer is yes, don't write anything. If the answer is no, list the name of the person who owns the respirator and explain what the problem is.

Is the respirator stored in an air-tight container?
_____

Is the respirator visibly clean?
_____

Is the respirator free of cracks, dents, cuts, or broken straps?
_____

Is it easy to breathe through the respirator?
_____

Is the respirator free of solvent odor?
_____

Are the prefilter and cartridge in good working condition?
_____

Summary:

I answered "yes" to every question for ___ respirators inspected.

The primary problem(s) with respirator care and maintenance is (are):
_____
_____

Inspector's Signature:_____

Figure 4-2. Respirator Inspection Form.

The company's "policy" regarding the inspection of respirators by both employees and managers should be clearly stated in the written Respiratory Protection Program.

Sample program excerpt: "All employees at XYZ Company who are assigned respirators are instructed at the time they are given the respirator, and at least annually thereafter, to inspect the respirator for defects and to conduct a leak check before each use. Employees are instructed never to wear a malfunctioning respirator and to report the malfunction to their supervisor immediately. Employees at XYZ Company are not penalized for reporting malfunctioning protective equipment."

Sample program excerpt: "The Safety Manager at XYZ Company conducts unannounced monthly inspections of all respirators in routine use by employees or stored for emergencies. Dated reports for each inspection are kept in a file in the Safety Manager's office and are reviewed at monthly safety meetings with production personnel."

2. Repair—When defects or malfunctions are discovered, repairs or adjustments must only be made within the *manufacturer's recommendations.* Replacement parts must be those designed for the respirator on which they are being used. This is critical where self-contained breathing apparatus (SCBA) are concerned. Components are not interchangeable among manufacturers. A North canister should not be used with a 3M mask or a Scott regulator.

   Trained technicians must conduct repairs on respirators of any type. Where instruction booklets indicate that certain parts, such as reducing or admission valves or regulators, should be returned to the manufacturer for repair, then by all means return them to the manufacturer. Those little instruction booklets provide a great deal of valuable information, much of which can be easily adapted for employee training. Don't throw them away. Read and follow them.

3. Cleaning—Although it might seem logical that the cleanliness of a piece of neoprene that a person has pressed against his or her face all day would be of great interest and concern to that person, without any prodding or reminding on the part of managers, this is not always the case. Sometimes, unless an employee has been shown how to clean the respirator and been told to do it regularly, it simply won't occur to him or her.

   The regulation states that routinely used respirators must be collected, cleaned, and disinfected as often as necessary to insure that proper protection is provided for the wearer. How you decide to handle this at your company will depend upon several factors, including the "dirtiness" of your operation and the diligence of your personnel.

   You might establish a routine where personnel who wear respirators bring them to weekly or monthly safety meetings or other production meetings. At that time all respirators can be inspected, thoroughly cleaned with soap and water (alcohol just dries out the facepiece and causes it to wear out sooner), and air dried. If regular production meetings are impractical at your company, you yourself can collect the respirators used on each shift (don't forget night shift employees!), clean them, and return them to their owners. If there has never been a regular schedule for cleaning respirators at your company, it might take a period of careful observation on your part to determine how often this needs to be done. Overkill is certainly unnecessary and will only serve to burn you out.

As an aid to figuring out an appropriate cleaning schedule, you might want to include cleanliness as part of your routine inspection for defects and proper fit. If you decide to keep written records of inspections, you certainly can include space for comments on cleanliness, including the condition of the inside of the facepiece and the cartridges. (Take another look at Figure 4-2.) This is also a clue as to whether or not the employee has been wearing the respirator and storing it properly. For instance, if the inside of a respirator is thick with dust, this is pretty conclusive evidence that, not only is it not being stored properly, but it's not being worn either.

The goal of your cleanliness program should be to make cleaning respirators a *habit*, not a hassle. Most employees would think it very strange to wear the same work clothes over and over again, without cleaning them. Wearing a dirty respirator should and *can* become just as strange to them.

Sample program excerpt: "Here at Smith Electroplating Company, employees who wear air-purifying respirators bring them to the monthly production meeting, where they are inspected and cleaned under the supervision of the Safety Manager. When the Safety Manager conducts random inspections of routinely used respirators, he checks for cleanliness as well as defects. If he discovers an exceptionally dirty respirator, he brings it to the immediate attention of the employee who owns the respirator."

4. Storage—To store a respirator improperly is to throw money out the window. A respirator that is not stored in such a way that it is protected against dust, chemicals, sunlight, excessive heat or cold, or moisture will wear out long before its time. But even more significant than the monetary loss, when respirators aren't properly stored, the whole Respiratory Protection Program breaks down.

Imagine that a painter has been told that a respirator should last six weeks if used daily in the work environment. But the painter "stores" his respirator simply by setting it on the paint mixing table whenever it isn't being used. As the respirator sits on the table, the cartridges absorb whatever solvent vapors are present in the shop, and dust coats the facepiece and settles in the valves. Under these conditions the respirator functions efficiently enough to protect the painter for only about two and a half weeks before the cartridges need to be replaced. So, three and a half out of every six weeks, this painter is *not adequately protected* from the respiratory hazards in his or her workarea, even though he or she wears the respirator diligently. And all because no one told the painter *how* to store the respirator to make it last.

The examples of this problem are literally endless. A truck driver who transports hazardous materials keeps his respirator tucked up on the dash board, in easy reach. Day after day the mask bakes in the sun, losing its proper shape. And day after day the cartridges absorb the smoke from the driver's cigarettes. If the driver needs the respirator in an emergency, what kind of condition will it be in?

Improperly stored respirators are dangerous because they give their users a false sense of security. They become dirty quicker than respirators that are properly stored and therefore are less inviting to wear. And if discovered, improperly stored respirators cost employers more dollars because they have to be replaced more often than respirators that are properly stored.

So what is proper storage? Proper storage means keeping the respirator away from anything that will damage it while it is not on the user's face. As with every other part of your Respiratory Protection Program, how you decide to handle this at your company is strictly up to you.

Dust masks or half-mask respirators can be put in zip-loc bags, as long as the bags are, in fact, zipped. Tupperware or some other type of air tight plastic box is effective for larger respirators. Merely sticking a respirator in a locker is not sufficient, especially if work clothes, boots, or gloves that may be contaminated with chemicals are also stored in the locker. By the same token, tossing a respirator in a tool box or drawer or cramming it under a car or truck seat is not good storage practice either. In addition to being exposed to air contaminants, these respirators are likely to become damaged from rough treatment or from being kept in an abnormal position. However respirators are stored, it is important that the facepiece and exhalation valve are allowed to rest in a normal position so that neither the function nor fit of the respirator will be impaired.

Respirator storage should be an important part of every training class where respirators are discussed. Management inspections of both routinely used and emergency respirators should always include checking for proper storage practices. As with the wearing of respirators, employers may find it necessary to implement a policy of disciplinary action for failure to properly store respirators.

Sample program excerpt: "All painters at ABC Body Shop are given air-tight plastic boxes in which to store their respirators. Bodymen are given zip-loc bags for storage of their dust masks. During routine respirator inspections, the Safety Manager checks to see if any respirator not in use is stored properly. Inspections are recorded by the Safety Manager. Improperly stored respirators are brought to the immediate attention of the employee who owns the respirator. Because proper storage of respirators is essential to the safety of ABC personnel, we have implemented the following policy: The first time an employee is found storing a respirator improperly, he or she is given a verbal reminder. The second time, he or she is given a written reminder. The third time, he or she is suspended from work for three days. The fourth time, he or she is dismissed."

Obviously, this chapter has not provided a comprehensive review of protective equipment. What the chapter has provided is detailed information regarding how to interpret the specific Standards covering the types of protective equipment that should be used while handling hazardous materials and wastes. However, if you are required to develop a program under a different regulation, such as Title 29 CFR Part 1910.95 on Hearing Conservation, you will find that the same site-specific, need-based approach to developing the program and training your employees will work for that Standard as well.

5
Plan, Prepare, Prevent

*Our plans miscarry because they have no aim. When a
man does not know what harbor he is making for,
no wind is the right wind.*

                              Seneca

Although nearly every safety and environmental regulation intersects with the Right-to-Know Law or Hazard Communication Standard at some level, it would be terribly confusing to attempt to analyze of all of them in one book. The OSHA Standards discussed here and in the previous chapter have been selected first, because their application is as universal as that of the Hazard Communication Standard itself, and second, because they offer an excellent illustration of three concepts essential to employee safety under normal working conditions as well as emergencies:

Planning for safety and health by developing sound procedures.
Preparing for safety and health by providing proper equipment.
Preventing accidents and emergencies by conducting training and inspections.

    As discussed in Chapter 4, the OSHA Standards for Personal Protective Equipment deal with the employer's responsibility to provide proper equipment for protecting employees from whatever workplace hazards exist. But more than that, these Standards emphasize the importance of using operating procedures, training, and inspections in order to prepare for and thus prevent emergencies. Where protective equipment is concerned, an emergency can range from an employee getting a foreign object in his or her eye to respirator failure in an IDLH (immediately dangerous to life and health) atmosphere. In all cases, preparation is the best method of emergency prevention.
    Compliance with these Standards is essential as part of the basic structure of any training program. Why? Because there is no safety or environmental regulation that deals with hazardous substances that does not intersect with the OSHA Standards for Personal Protective Equipment. You can't comply with OSHA's

Hazardous Waste Operations and Emergency Response Standard, Title 29 CFR Part 1910.120 without addressing protective equipment. You can't comply with RCRA's Hazardous Waste Generator requirements for Contingency Planning, Preparedness and Prevention, or Personnel Training without detailed coverage of protective equipment. You can't provide adequate training under DOT for drivers who transport hazardous materials or hazardous waste without, once again, dealing with protective equipment.

OSHA's Standard covering Employee Emergency Plans and Fire Prevention Plans outlines, in very practical terms, the plans and procedures that employers must implement in order to minimize the impact of emergencies or avoid them entirely. Emergencies covered by this standard include fires, spills, and natural disasters.

Every employer must have an Employee Emergency Plan and Fire Prevention Plan. However, if you happen to be a small- or large-quantity generator of hazardous waste, or a treatment, storage, and disposal (TSD) facility, this plan can also function as the basis of (or even stand in place of) the Contingency Plan and Emergency Procedures and Preparedness and Prevention Plan required by RCRA. What is more, if you are a small- or large-quantity generator or TSD and your employees are *not trained to respond* to emergencies, but *to evacuate,* then you are exempt from compliance with OSHA Title 29 CFR Part 1910.120, as long as you have an Employee Emergency Plan and Fire Prevention Plan and you train your employees accordingly. A well thought out plan is not only an aid to your employees, but it can help you streamline your compliance. (See Chapter 8 for more suggestions on combining training requirements.)

Lost in the shadow of the "big" emergency planning regulations, OSHA's Hazardous Waste Operations and Emergency Response (HAZWOPER), Title 29 CFR Part 1910.120, and RCRA's Contingency Plan and Emergency Procedures, Preparedness and Prevention Plan, and Personnel Training Requirements, Title 40 CFR Part 264.16 and Subparts C and D, this little Standard (it takes up less than a page in the CFR) is a sleeping giant.

If you've never heard of 1910.38, don't worry. You're not alone. But once you get to know this regulation you're going to want to kick yourself for not hearing about it sooner. Of the key safety regulations impacting your business, this one provides the most practical and thorough guidance. This one really does make sense.

Regardless of the type of operation you have, whether or not you generate hazardous waste or have an emergency response team, complying with this regulation *first* will make complying with other emergency planning requirements worlds easier. You can't develop a meaningful Employee Emergency Plan without Hazard Communication and Personal Protective Equipment programs as your foundation. But once you've integrated these three regulations into the management of your company, you'll find that complying with nearly any other training requirement is just a detail—the equivalent of furnishing a house after the hard work of renovating the kitchen, painting the walls, and laying the carpet is all over with. (See Chapter 8 for more on combining compliance requirements.)

While we might think of emergency and fire prevention plans as being more critical at an industrial facility, where lots of hazardous materials are present, they are necessary any place employees work where there is the possibility of fire, chemical spill, or natural disaster, which is, of course, *everywhere*. It is critical that employers at industrial facilities remember that it is not just production personnel who need to be included in emergency planning. Often, employees who work in office and other non-production-oriented areas are completely left out of planning, training, and emergency drills. In an emergency, office personnel may well be in as great a risk as people on the production floor; it is essential that they know how to evacuate and where to rally.

Additionally, when non-production personnel are not involved in planning and training, employers miss an opportunity to involve them in the *emergency management* process. For instance, office personnel can serve in valuable roles during emergencies in communicating with outside services, making public address announcements, triggering alarms, picking up time cards or a visitor's registry during an evacuation, and taking roll at a rally point.

What Does OSHA's Standard for Employee Emergency Plans and Fire Prevention Plans Require?

The requirements of this Standard, as stated in the regulation, are very straightforward and demand little translation. However, as an aid to those employers who have never developed an emergency plan or are having difficulty getting started, many examples have been included here showing how these requirements have been met at various workplaces. Additionally, where a requirement of this Standard directly duplicates or intersects with a requirement of RCRA's Hazardous Waste Generator regulations found in Title 40 CFR Part 264, including Contingency Plan and Emergency Procedures, Preparedness and Prevention Planning, or Personnel Training, the duplication or intersection is noted.

It is important to mention that the primary reason for not offering an analysis of Title 40 CFR Part 264 in this book is that, due to the open-ended language of the regulation, there is currently no consensus of opinion regarding the contents of a training program under this regulation. There is no one training program that is easily recognizable as an RCRA training program. At present, trainers across the country are working with associations such as the American Society of Testing and Materials (ASTM) to develop a standard content guide for training required under RCRA. Once such a standard exists, it will be possible, in another book, to break it down and identify how its requirements can be met at various workplaces.

EMPLOYEE EMERGENCY PLANS

Regulatory Code:

> 1910.38(a) Emergency Action Plan (1) Scope and Application. The Emergency Action Plan shall be in writing (except where employers have 10 or fewer employees; for those employers the plan may be communicated orally to employees) and shall cover those designated actions employers and employees must take to ensure employee safety from fire and other emergencies. (2) Elements. The following elements, at a minimum, shall be included in the plan: (i) Emergency escape procedures and emergency escape route assignments. (ii) Procedures to be followed by employees who remain to operate critical plan operations before they evacuate; (iii) Procedures to account for all employees after emergency evacuation has been completed; (iv) Rescue and medical duties for those employees who are to perform them; (v) The preferred means of reporting fires and other emergencies; and (vi) The names or regular job titles of persons or departments that can be contacted for further information or explanation of duties under the plan. (3) Alarm System. (i) The employer shall establish an employee alarm system which complies with 1910.165. (ii) If the employee alarm system is used for alerting fire brigade members, or for other purposes, a distinctive signal shall be used for each purpose. (4) Evacuation. The employer shall establish in the emergency action plan the types of evacuation to be used in emergency circumstances.

English: What is the purpose of the Emergency Action Plan? To maximize employee safety and health in an emergency. An emergency, as defined by this Standard, includes a fire, chemical spill or release, hurricane, tornado, blizzard, flood, or other natural disaster.

Developing a Plan

There are at least five areas that must be covered in your Emergency Action Plan: Communications and Alarm Systems, Evacuation Procedures, Shutdown of Operations, Rescue Procedures, and Leadership.

Communications and Alarm Systems

Your Plan needs to include the methods that employees are supposed to use to report or notify each other of emergency situations. According to Title 29 CFR Part 1910.165, the method of reporting can range from pulling a box alarm to accessing the public address system or using a radio, walkie-talkie, or telephone. At workplaces with ten or fewer employees, direct voice communication is an acceptable procedure for sounding the alarm. Regardless of the method used, however, employers must verify that employees can in fact hear the alarm. If ambient noise levels inhibit the employees' ability to hear the alarm, then a backup visual alarm must be used.

Case in Point: At a chemical manufacturing company, a great deal of internal construction has been completed since the public address system was installed. Several mixing and process rooms in the center of the building do not have their own speaker. During an emergency drill it is discovered that employees working in these rooms cannot hear emergency announcements made over the public address. As a result of the drill, company management takes steps to install additional speakers in these rooms.

Where communications systems, such as telephones, PA systems, or radios, also serve as the alarm system, it is a good idea to establish some sort of code or signal whereby employees will immediately recognize whatever is said as relating to an emergency. Also, it is helpful to post important emergency telephone numbers, the telephone extensions of key personnel, public address access numbers, or any other special code numbers or communication procedures on telephones and bulletin boards, or even to print them on wallet cards that employees can carry with them. Precisely how you choose to manage the reporting of emergencies at your company will depend on a variety of factors that must be taken into account in order for the system you choose to work.

Case in Point: A chrome furniture manufacturing company with 300 employees has 11 production departments spread out over several thousand square feet. Each department is referred to by a number, and the numbers are not in any kind of recognizable order because, as the company has grown and changed, departments have come and gone, taking their numbers with them. Consequently, the powder painting department is number 31, while the wet paint line right next to it is number 12, and so on throughout the plant. Since no one could remember which number went with which department, the company printed up laminated wallet cards for designated emergency personnel in each department. When there is a problem in a department, the designated employees have been instructed to go to a telephone, punch in a number that accesses the PA, punch in another number that causes three loud beeps to come over the PA, and then make an announcement. If there is a fire, they announce: "Department 31, Code Red." If there is a spill, they announce: "Department 31, Code Blue." If there is an employee injury, they announce: "Department 31, Code Green." If evacuation is necessary, they announce: "Department 31, Evacuate North," or whichever direction the personnel need to go to avoid the emergency. The wallet cards include an explanation of the codes. Depending on the code announced, the designated emergency personnel have been trained either to go to Department 31 or to begin assisting the personnel in their department in evacuating the building.

Different codes or announcement procedures might need to be developed for natural disasters. In areas where tornadoes, hurricanes, or floods are likely to occur, employers may want to establish a warning horn, siren, or succession of sounds over the public address that would serve as an indicator to all personnel to take cover in whatever manner is best suited to the emergency at hand. But remember, if employees don't understand the alarms, or if they don't know how to respond when they hear them, the system will fail. On-going training is essential for such a system to work. For instance, when a tornado is sighted, an employer sounds a particular siren that all employees have been trained to recognize as a tornado warning. In accordance with their training, each employee stops work, turns off their equipment, if it is possible to do so, and proceeds to the lower level of the building, which has been designated as a shelter.

It is a fact of life that you will never know how well a communications and alarm system does or does not work until you try it. And the preferred time to try it and work out the glitches that are bound to exist is *well before an actual emergency.*

The requirements of RCRA's Part 264.16 on Personnel Training, and Subparts C and D on Contingency Plan and Emergency Procedures, and Preparedness and Prevention Planning also specify the need for communications and alarm systems, although they do not provide as much guidance as OSHA's Standards as to what those systems can or should be. RCRA requires that an internal communication or alarm system be available to provide immediate emergency instruction to facility personnel and that all personnel know how to use communications and alarm systems or recognize signals or instructions received through them.

Evacuation Procedures

The reason for developing evacuation procedures is to figure out exactly what route each employee is supposed to take to get out of the building as quickly and safely as possible. General evacuation procedures is an oxymoron, a contradiction in terms. Evacuation procedures must be specific, not only to your facility, but to *each employee,* in order to be meaningful.

Obviously, if your company is similar to that of the furniture manufacturer outlined above, your evacuation procedures will have to be quite detailed, not only in terms of where the personnel in each department need to go, but in terms of how their routes might change given the type and location of the emergency.

In contrast, if you run a machine shop where all 20 of your employees are in one room and there are only three exits, then your evacuation plan will be quite simple, because there are only three ways for your people to go, regardless of the type of emergency. Regardless of the size of your facility, the number of employees you have, or the relative simplicity or complication of your evacuation plan, you must *map it out.*

If you don't have a drawing of the workplace and don't have anyone on staff who can make one, try to track down the construction blueprints or floorplan. A blueprint supply company can copy and reduce the size of your prints. On these you can indicate evacuation routes by department, if you have a large facility, or even for each individual or work station, if that would work better. You might want to color code routes for departments or individuals. Or, rather than making one evacuation map for the entire facility, you might want to make several maps, each showing only the evacuation route for a particular department. At some facilities it works better to design route maps for each employee and post a map at each work station.

The important thing to remember is that evacuation plans and route maps won't do anyone at your company any good unless they know about them and, they practice them. Posting the route maps is important, and it is a requirement mirrored in RCRA's Contingency Planning requirement, but if your employees have never physically evacuated the building during an emergency drill, then they won't know how to get out in a real emergency, even if they've been staring at a route map for years.

The evacuation plan doesn't end when everyone starts evacuating. True, the purpose of the evacuation is to get everybody out. But how do you know they're out? You must have a procedure for accounting for everyone who was in the building when the emergency occurred. At some companies, this is called "Roll Call," which is a good name for it because it implies that someone has a *roll,* or a list of people who were in the building, to *call* from.

Remember fire drills in elementary school? Everyone filed out of the classroom, down the hall, out the nearest door, and onto the playground by the swing set, or to some other specially designated spot, where the teacher proceeded to call roll. The procedure you need to develop for your company is no more complicated than the one you followed in third grade. It is, however, often quite a bit more difficult, since, unlike your teacher, you probably don't have an attendance record in your desk drawer. So what can you do?

First, you have to figure out who at your company knows who is or is not at work on any given day. If you run a small printing company, whoever is in charge of the front office probably knows. They will be the person everybody checks in and out with, the person who knows who had a dentist appointment and who had to go home because their kid has the flu. They will know if a vendor is on-site filling up the Coke machine or repairing the air conditioner.

If your operation is a bit more complicated than this, you will probably have to delegate the responsibility for knowing who is and isn't at work among several individuals. If you have multiple departments, the supervisor or head of each one can be given the responsibility both for knowing who is and isn't at work (including whether or not contract employees or vendors are present) and for being the last one out of the area when an evacuation is called. The supervisor can conduct a quick visual check through the department as he or she leaves, including taking a look in breakrooms and restrooms. When the supervisor and the department personnel reach the designated rally point, he or she can check in with the designated Roll Call Officer to report who is present and who is unaccounted for. The Roll Call Officer will then report this information to whoever is in a position of leadership on the scene, such as the Emergency Coordinator (a role required by RCRA) or Fire Chief.

Evacuation maps always should indicate the location of the rally point or points where personnel are supposed to gather when they evacuate. The rally point always should be distant enough from the building to insure that personnel will be reasonably protected in the event of spill or fire. It also should be out of the path of any emergency service vehicles that might respond to an emergency. Don't have your personnel rally in the parking lot if that is where the fire trucks will have to set up to get to your chemical storage. Instead of setting a rally point in the middle of the street, why not have it across the street, on the lawn in front of another building? Just to be sure this doesn't get you in any legal trouble, ask the owner of the property for permission to use the area as a rally point. If he or she agrees, get a written letter saying as much. Whether you are across the street from another business, a homeowner, or a vacant lot, most folks will be pleased and relieved to discover that you have an active emergency planning program and they will be happy to help.

In order to be valuable as a training tool, evacuation procedures must be built around emergencies that either have occurred in the past or might conceivably occur. In short, they must be *realistic*. Actually, coming up with evacuation scenarios can be a lot of fun, especially if you have a latent creative streak or you enjoy problem-solving. So how does this work?

First, you don't look at your whole facility at one time. Regardless of its size, you have to take one department, one area, or even one piece of equipment at a time and ask yourself as many "what if" questions as you can dream up. If you don't have an in-depth understanding of production operations, you might want to go out to the department supervisor or equipment operator and ask him or her the "what if" questions. What if a spark was created in a chemical storage area? What if a forklift caused a puncture in a drum? What if a valve malfunctioned on an acid tote? What if a transfer hose on a fuel tanker ruptured?

The Material Safety Data Sheets for any hazardous materials used in a particular department or piece of equipment can provide a starting place for answering these questions.

Case in Point: Greg, the Safety Manager at an electroplating company, is developing evacuation procedures for a plating line. At one place along the line there is a nitric acid tote. In order to figure out when evacuation might be necessary, he needs to determine what could happen if the nitric acid was spilled. Going to the the Material Safety Data Sheet, he finds out that nitric acid is an oxidizer and that if it comes in contact with combustible materials, such as the wood pallets stored just 6 feet from the tote, highly flammable hydrogen gas will be created. The Material Safety Data Sheet also tells him that nitric acid fumes are highly toxic, allowing for a very short exposure period. From this information Greg concludes that nitric acid spills or leaks would require the immediate evacuation of personnel and the notification of emergency services. When he conducts training for personnel who work along the plating line, he includes discussion of the specific hazards created by nitric acid spills and the need for immediate evacuation whenever they occur. He even goes so far as to run a drill by spilling water along the plating line that he tells the employees to respond to as if it were nitric acid.

The Contingency Plan required under RCRA must include the same kind of written evacuation plan required by this Standard. Title 40 CFR Part 264.52(f) says that the "plan must describe signal(s) to be used to begin evacuation, evacuation routes and alternate evacuation routes (in cases where primary routes could be blocked by releases of hazardous waste or fires)." If you already have an evacuation plan in your Contingency Plan, with perhaps slight expansion to include sources of releases or fires other than hazardous waste, this part of your RCRA compliance can double as your OSHA evacuation plan. The reverse may also be true. (For more on combining compliance and training requirements, See Chapter 8.)

Procedures for Shutdown of Operations
The Emergency Action Plan must include a detailed explanation of who will be responsible for maintaining or shutting down essential plant operations in the event of an evacuation and how they will go about doing it. Essential plant operations might include power supplies, gas or water supplies, or chemical or manufacturing processes that must be shut down in stages.

Case in Point: The maintenance staff at the same plating company mentioned above is designated to remain behind to monitor or shut down operations in the event of an emergency evacuation. In order to insure that the workers are in an optimum position to protect themselves from whatever emergency is present, they have each been given a walkie-talkie, so that they can maintain communication with each other and with the designated Emergency Coordinator. They also have been given respirators, goggles, gloves, boots, and protective clothing, which they keep near them in their maintenance carts. It is not intended for these personnel to put their lives at risk in the process of shutting down or monitoring operations. When it is relatively safe to do so, they are instructed, in the case of fire, to turn off gas lines, turn on alarm and sprinkler systems, and monitor the shut down of process equipment to make sure it doesn't make the situation worse. In the case of a spill or release, they are instructed to turn off any equipment causing the release, to eliminate ignition sources, and to cut off power supplies as necessary.

RCRA's Personnel Training requirements found in Title 40 CFR Part 264.16(3) also require designated personnel to be trained to be able to cut off automatic feed systems and conduct a successful emergency shutdown of operations.

Rescue Procedures

Assignment of medical or first aid duties to specific personnel is certainly optional. It may be that you have no one at your facility who is capable of performing first aid or CPR. If this is the case, then, in the event of a medical emergency you would instruct your personnel to call for help. There is *no* company, however, that will not benefit from having personnel trained to perform first aid and CPR. Paying for this training may in fact be one of the best investments an employer can make in the safety of everyone at the company.

If you decide to train some or all of your personnel in first aid and CPR, then they must receive annual update training to maintain their certification. This should be outlined in the Emergency Action Plan and explained during your emergency training. You may decide to ask for volunteers to become certified. Before they volunteer, your employees need to be aware of the time involved in certification, as well as of the duties they may be expected to perform during on-site emergencies.

The Plan should include the names of the personnel who can provide first aid and CPR and where or to whom they should report during an emergency. In order to avoid arguments about who is going to do what at the scene of an emergency, the division of labor should be fairly specific.

Sample program excerpt: "John Smith will bring first aid and CPR supplies and equipment to the scene of the emergency or the rally point, as instructed by the Emergency Coordinator. Harry Davis will call for outside emergency medical services. David Brown will conduct CPR. Sue Williams will treat burns. Steve White will treat cuts. Gail Baker will treat broken bones."

In the Plan you might want to describe, and thus limit, the "treatment" that rescue employees are qualified to give.

Sample program excerpt: "In the case of a suspected broken bone, the rescue employee's job is limited to monitoring the injured employee, and making sure he or she remains still, warm, and stable."

The purpose of first aid and CPR is to keep an injured person alive and to prevent their injuries from worsening *until trained medical personnel arrive on the scene.* The purpose is not to "play doctor." The difference must be clearly explained to your rescue personnel.

If you operate more than one shift, don't forget the need for rescue volunteers from each shift. It is also necessary to have an alternate for every rescue position on each shift. All of these employees must participate not only in annual first aid and CPR certification, but also in on-site training and emergency drills. Only after repeated practice of their skills in on-site drills will they be competent and capable enough to apply those skills in a real emergency.

Leadership

For lack of a better term, the title of Emergency Coordinator has been used several times throughout this analysis to indicate "the person in charge." Naming an Emergency Coordinator is part of RCRA's Contingency Planning requirements. The term also comes up in OSHA 1910.120. However, the term itself is not used in this Standard, 1910.38, which simply requires the "names or titles of the people who can be contacted for more information or for an explanation of duties under the Plan."

This would seem to imply, however, that one or more people at a company really need to understand the ins and outs of the Emergency Action Plan and have ultimate authority in implementing it. At least one person needs to be the leader. When the evacuation has been called and the fire department shows up, somebody needs to step forward who can tell the Fire Chief what is going on and whether or not anyone is left inside. When the maintenance crew stays behind to shut down the plant, they need someone on the other end of the radio to check in with, someone to tell them what to do.

Without someone in the lead, it is unlikely, if not impossible, that you will have a successful emergency response, even if that response is limited to evacuation and shut down. You can call this person an Emergency Action Leader, or a Chief Emergency Officer, or any other name you can come up with, but regardless what you call them, they will be filling the same role as that of an Emergency Coordinator.

The Emergency Action Plan should describe the duties of the Emergency Coordinator, along with those of all other specially designated emergency personnel who make up your Emergency Action Team. As mentioned throughout this analysis, team members might include:

Supervisors: These personnel serve a valuable role in activating communication and alarm systems when emergencies occur, in directing the personnel in their department to evacuate, in checking to make sure everyone in their area has evacuated, and in reporting any missing personnel to the person taking roll at the rally point or to the Emergency Coordinator.
Maintenance Crew: These personnel remain behind in an evacuation, either to monitor essential plant operations and process equipment or to shut it down in such a manner as to avoid further problems.
Rescue Personnel: These employees are certified in first aid and CPR and have received on-site training in their specific roles in providing limited treatment to injured personnel.
Roll Call or Communications Personnel: These personnel are responsible for determining whether or not anyone remains inside the building after an evacuation and for reporting the information to the Emergency Coordinator or Fire Chief. They may also be directed by the Emergency Coordinator to contact emergency services or to maintain radio contact with maintenance personnel remaining inside the building.

Regardless of the size of your company, you need to think through exactly how these roles are going to be filled. As you do this, several questions undoubtedly will arise.

Who would be the best Emergency Coordinator?

Whoever it is, just remember that he or she needs to know a lot about the operation, the chemicals and hazards that are present, and the Emergency Action Plan. The company President, Environmental Compliance Manager, or even the Safety Manager may not be the best choice for the job—at least not if they are removed from ongoing contact with facility operations. A better choice might be the Plant Manager, Production Supervisor, or head of Maintenance.

Do I have to involve my supervisors?

Yes. Whether you have a one-room printing shop with 8 employees or a manufacturing company with 300, you need to have personnel out in the production area (and in non-production areas as well) who will take responsibility for seeing that everyone evacuates safely. At the one-room shop, you only need to involve one supervisor. At the manufacturing company, you may need to involve 15 or 20. One rule of thumb: One well-trained supervisor for every 20 employees is needed to lead a safe evacuation.

Does someone have to stay behind to shut everything down?

Actually, no. If you feel more comfortable having a policy that when an evacuation is called everybody leaves, that's fine. Establish the policy, put it in your Emergency Action Plan, and train your personnel accordingly. If, however, you decide to implement procedures for emergency shut down, and you have a maintenance crew, they are probably the best folks for the job. If you don't have a maintenance crew, department supervisors may be able to shut down equipment and utilities as they follow behind their evacuating personnel. Just remember that whatever you decide to do, you need to fully explain both the procedure and who is going to carry it out in your Emergency Action Plan.

What do Rescue Personnel have to know?

As mentioned above, having rescue personnel is not mandatory, although it is recommended. If you don't want any of your employees to perform first aid or CPR, then simply say so in your Plan. If you do choose to have rescue personnel, however, they should be volunteers who have completed the required course work and been certified to perform first aid or CPR by an agency such as the Red Cross.

Who should be in charge of Communications or Roll Call?

Depending on the size of your organization and the way it is set up, personnel with a variety of different job titles may function well in this role. In order to improve the odds of getting someone who is committed to the job, it should be filled by a volunteer. However, a front office employee is often a good person to encourage, since they usually have access to a visitor's log and may have access to a log where employees or contractors sign in and out throughout the day. When an evacuation is called, this employee can pick up the available logs en route to the rally point. Additionally, this person can aid the Emergency Coordinator by making emergency announcements over the public address or maintaining contact with personnel conducting emergency shutdown. Or a system can be set up where, in response to an announcement or to an alarm, this person calls outside emergency services immediately before evacuating the building.

Again, the reason for spelling out different Emergency Action Team roles is not to limit you to one approach in handling emergencies and evacuations at your company, but to emphasize the absolute necessity of designing a Plan that is practical and of filling each role in the Plan with the personnel who are most capable and willing to do the job (see Figure 5-1).

Regulatory Code:

1910.38(a)(5) *Training*. (i) Before implementing the Emergency Action Plan, the employer shall designate and train a sufficient number of persons to assist in the

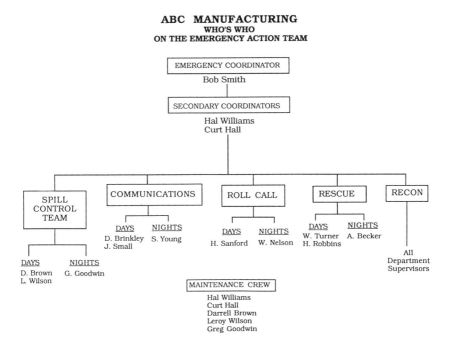

Figure 5-1. Who's Who on the Emergency Action Team.

safe and orderly emergency evacuation of employees. (ii) The employer shall review the Plan with each employee covered by the Plan at the following times: (A) Initially when the Plan is developed. (B) Whenever the employee's responsibilities or designated actions under the Plan change, and (C) Whenever the Plan is changed. (iii) The employer shall review with each employee upon initial assignment those parts of the Plan which the employee must know to protect the employee in the event of an emergency. The written plan shall be kept at the workplace and made available for employee review.

English: Who has to be trained?

The Standard is very specific in directing the employer "to designate and train a sufficient number of persons to assist in the safe and orderly evacuation of employees." This refers to the Emergency Action Team members with the specific task of helping other employees get out of the building, or, as defined above, the Emergency Coordinator and Supervisors.

How do you determine a "sufficient number" ? It is based upon the size of your operation. If you have a shop with 8 employees, you only need 1 Emergency Coordinator and 1 Supervisor. However, if you have a company with several hundred employees, you will need 1 Emergency Coordinator and perhaps 20 Supervisors.

Training Leaders to Lead

What do these people need to know about the Plan? Is it enough for them to know the contents of the Plan, the five sections listed above? No, knowledge of procedures and policies alone isn't enough. The Emergency Coordinator and the Supervisors must be able to decide *when to use them*. Their first responsibility is *to recognize* emergency situations that require evacuation. Their second duty is to know enough about the communications and alarm systems, the facility layout, the hazards of the workplace, and the evacuation routes, *to lead* other employees to safety.

As with the Hazard Communication Standard, the Personal Protective Equipment Standards, and RCRA's Personnel Training, OSHA does not require training to be of a certain length. It is up to the employer to create a training program that will sufficiently prepare employees to lead a successful evacuation. Therein lies a sneaky little catch: If your employees fail to evacuate safely in a real emergency, then the burden of proof returns to you, the employer, to prove that enough training was provided that they should have known better—that is, that their failure to evacuate was as a result of *their* negligence, not *yours*. Which may be difficult to prove indeed.

So how are you going to insure that your Emergency Coordinator and Supervisors are themselves adequately trained to lead a successful evacuation in a real emergency?

You may be thinking that there is not enough time or money in the world to insure that. Take heart. It's not that bad.

Pick Their Brains

What you have to do is get the Emergency Coordinator and Supervisors involved in the *development* of the Emergency Action Plan. If you get them involved on the front end, not only will the Plan be more practical, but they will end up actually training themselves (and teaching you a few things in the process). Yes!

As the person responsible for their training, don't *tell* them what an emergency is and what to do when one occurs, *ask them.* Ask them pertinent questions regarding the elements of emergency evacuation that must be covered in the Plan. And then *allow them to figure out the answers.* Pretend you're the captain of a ship. Whereas you chart the course and you point the ship in the right direction, it is your crew at the oars that provides the muscle to get the ship from one point to the next.

This doesn't mean that you should sit back and let your class do all the talking. Don't kid yourself into thinking this is the easy way out. Quite the contrary, keeping the class on course, moving steadily toward the goal of creating a useful Emergency Action Plan is about as far from easy as the sun is from earth. Posing realistic "what if" questions, leading and directing discussion without being intrusive, allowing the people you're training to feel that they have all the answers, and leaving your ego at the door is a very difficult form of training. Much more difficult than a didactic lecture, and much more effective.

Case in Point: Dennis is the Safety Manager at A-1 Manufacturing. He is conducting one of a series of weekly meetings with Emergency Action Team Supervisors responsible for leading their personnel in evacuation. At this meeting, they are working on developing evacuation procedures for each department. The procedures will be included in the Emergency Action Plan for the company.

As a "reality check," Dennis has developed a feasible emergency scenario for each department. It is around these scenarios that they are building the discussion.

"Okay, folks, here it is: There is a leak in one of the hoses in the paint kitchen in Department 31. About 20 gallons of flammable solvent has leaked onto the ground, run underneath the door, and is accumulating around one partial and one full drum of flammable, paint-related hazardous waste sitting next to the door. This happens to be one of those unlucky days that somebody forgot to put the bung back on the waste accumulation drum. Instead of the bung being screwed in place, the funnel is sitting balanced in the bung opening. About this time, Jerome drives by on the forklift. Since we've all gotten a little lax on the forklift inspections, it turns out that the propane tank on the lift is leaking just a little bit. But enough, it turns out, to be ignited by Harold's cigarette when he stops Jerome to talk to him and leans against the tank with the hand holding the cigarette. Jerome and Harold both run away from the forklift. Neither one of them gets hurt. So what do you think might happen next?"

Based on this scenario, the Supervisors determined that an immediate evacuation of the building would be necessary. They also determined that Jerome and Harold, like all employees in non-supervisory positions, need to be instructed to report emergencies, such as fires and spills, to the nearest Supervisor they can find. The Supervisor they notified would then go to the nearest telephone, access the public address system, and make an evacuation announcement.

After further discussion, one of the Supervisors pointed out that without special instructions regarding where the emergency was occurring, they might actually lead their personnel into the heart of it. This prompted the Supervisors to develop an announcement format whereby they could indicate the safe direction in which to evacuate. They reviewed a map of the plant and figured out where personnel in each department should go in the event of the emergency Dennis described. They also determined how the routes might differ for the second shift if the same emergency were to occur at night.

Build on Procedures
As a spin-off from the discussion of evacuation procedures, the Supervisors began talking about the importance of maintenance, inspections, and standard operating procedures in keeping emergencies from happening in the first place.

One Supervisor says: "You know that whole disaster never would have occurred if people were doing their jobs properly. It's a scary thought, but every time one of us takes a short cut we're putting everybody else's life at risk. I know everybody in this room thinks it can't happen to him or her, but it can. You're in the wrong place at the right time and wham. It's too late to play catch-up then."

As they review the Safety Manager's scenario piece by piece, they realize that the fire and probable explosion were the direct result of four violations of what was *supposed* to be standard operating procedure all coming together at one time.

1. Failure to monitor the paint kitchen equipment.
2. Failure to inspect and repair the propane tank on the lift.
3. Failure to replace the bung on the hazardous waste drum.
4. Failure to stop smoking in the wrong area of the plant.

What more effective ways are there to underscore the importance of standard operating procedures in emergency prevention? Additionally, where standard operating procedures intersect with hazardous waste management, such as in the case of keeping the bung screwed in place on a drum of hazardous waste and moving a full drum of waste from an accumulation area to a storage area, it is easy to see the overlap between RCRA's Contingency/Preparedness and Prevention Plan and the Emergency Action Plan and between their respective training requirements (see Chapters 7 and 8). At least it is easy for these Supervisors to see, because they arrive at the conclusion themselves.

Clearly, the benefits to be gained by taking Dennis' approach to emergency training are enormous:

1. The Emergency Action Plan means much more to the Emergency Coordinator and Supervisors than "just one more regulatory paperwork requirement." Why? Because they helped write it. They put some of themselves into it. They're proud of it. And, by golly, they think it makes sense.
2. Standard operating procedures regarding maintenance and inspections, hazardous waste management, and smoking that the Supervisors themselves have skirted and at times even scoffed at will be followed with more regularity because they are connected, in the Supervisors' minds, to emergency prevention. Recognizing the crucial importance of these procedures, they will take a stronger role in making sure their personnel follow them.
3. The Supervisors will be enthusiastic about participating in evacuation drills because they will be curious to see if the routes and procedures they designed actually work. There may even be a certain amount of healthy competition stirred up between individuals with different ideas about what kind of alarm to use or where to establish a rally point.
4. They will pass on their enthusiasm to the employees they supervise. It will become a matter of personal pride to make sure that every individual in their department knows exactly what to do if an evacuation is called. They can be encouraged to talk about evacuation procedures routinely at any production or safety meetings they hold with their departments.

How many training sessions do you need to have with your Emergency Coordinator and the Supervisors who will lead the evacuation? You will need as many as it takes to develop a viable Plan and practice it until everybody can put it into effect. This isn't an evasion of the question; it is simply the only sensible approach to emergency training. You have to train until your Emergency Coordinator and Supervisors get it right. Then, after they get it right, you have to review it at regular intervals so that they don't forget.

The Emergency Action Plan is intended to be a living document. As discussed in Chapter 2, this means that it is not something you will write once and put on a shelf, never to be altered again. Through training and practice, flaws in even the most well thought out plans will be uncovered. Don't let that scare you. That is what is *supposed* to happen. Uncovering weaknesses or loopholes in an Evacuation Plan simply indicates that you and your employees are doing your jobs. By the same token, when your process changes, when you add on to the building, rearrange departments, or increase or decrease the number of personnel working in a certain area, you need to call a meeting with your Emergency Coordinator and Supervisors, review the plan in light of these changes, and alter it accordingly.

Training the Rest of the Team

So is it just the Emergency Coordinator and Supervisors leading the evacuation who need training under the Plan? No. The Standard goes on to say that the employer has to review the Plan with each employee covered by the Plan.

What does "covered by the Plan" mean? You can assume that all of the members of your Emergency Action Team, including whoever has been designated to handle emergency shutdown, your rescue personnel, and your roll call or communications officers, are definitely covered by the Plan, since they are mentioned in it.

Once the Plan, including the roles of these team members, has been developed, at least in a general sense, it is time to sit down with them and go over their individual roles. The operative question that must be answered for each of these team members is: "What do I do when I hear the alarm to evacuate?" The more *specific* their duties are, the greater the likelihood that they will be able to successfully carry them out.

Although their roles basically will be written for them, in training these individuals it is important to allow them to feel that they *have a voice in improving them.* They need to know that nothing is engraved in stone. Remember, the Emergency Action Plan is intended as a living document. As such, all ideas and perspectives must be welcomed. For example, as you discuss procedures for shut down of plant operations, encourage the maintenance crew to be critical of the procedures that have been developed. Ask them to dissect them and to try to uncover weaknesses and flaws. Encourage them to come up with worst-case scenarios and see if the procedures still hold up. Like the Supervisors, these team members will take a greater interest in the Emergency Action Plan if they feel they had an opportunity to add their two cents to it.

And don't forget that if you decide to have rescue personnel they need to have first aid and/or CPR certification *as well* as on-site training in their duties under the Plan. The certification alone is not enough. They must know exactly how they are expected to use their skills in on-site emergencies.

Training Everyone Else

What about all the rest of your employees—everybody who doesn't have anything to do under the Emergency Action Plan except get out of the building? What do they need to know?

The basics.

Everyone employed at your company needs to know:

1. That an Emergency Action Plan exists.
2. To report fires, spills, and leaks to the nearest supervisor as soon as the problem is discovered.
3. What an evacuation alarm sounds like.
4. What the evacuation routes are and where the route maps are posted.
5. To follow their supervisor's directions in an evacuation.
6. To wait at the rally point until they have been dismissed.

Depending on the size of your company, you may want to handle this part of your emergency training in one of several ways. Whatever you do, remember to *document it*.

Company Meetings
You can hold a company-wide meeting where attendance is mandatory. At the meeting you can cover the basics, introduce the members of the Emergency Action Team, distribute route maps, and perhaps even have the company President say a few words about the company's commitment to emergency preparedness and prevention. The meeting might turn into a morale session.

Departmental Meetings
You can hold meetings by department, where each Supervisor takes the lead in explaining the evacuation procedures to his or her own personnel. This will help insure that the information covered is workarea-specific, and that problems regarding evacuation routes and equipment shutdown are handled by the people who encounter them. Also, it may help to strengthen team spirit within your departments, as employees realize the importance of working as a unit.

One-on-One Meetings
You can hold one-on-one meetings with individual employees. This might be especially effective if many of your personnel have poor literacy skills. These individuals are likely to learn very little in a classroom setting. One-on-one meetings are the most effective way to train employees at the time they are hired. In fact, covering the basics of the Emergency Action Plan can be incorporated into whatever new employee orientation program you already use.

In Praise of Drills
Although the Standard does not mention evacuation drills, they have come up multiple times throughout this analysis. Why? Because drills are literally the only way you can tell if your employees know how to follow the Plan and if it will work. In fact, a Plan that has not been practiced through emergency drills is worth about as much as the paper it is printed on, or even less. Because a written Plan that has been discussed in training sessions gives everyone a false sense of security, a false

102   Safety and Environmental Training

cockiness. Sure, we know what to do, everyone thinks. We've gone over it a million times.

Would you teach someone how to drive an automobile in a classroom setting and then, the first time you put them behind the wheel, expect them to drive an ambulance to the hospital at 60 mph through rush hour traffic? Of course not. That would be lunacy. The odds are that they would have a wreck. Telling your employees how to evacuate without having them practice evacuating is about as sensible. Under these circumstances, what are the odds that, in a real emergency, they will evacuate safely? *Drills are the insurance that the odds will be in your favor, and theirs.*

The essential nature of on-site drills is the most compelling reason why canned, prepackaged training programs and off-site training classes are anathema to safe, successful emergency response. Such canned programs may or may not serve to enhance your site-specific emergency training, but do not kid yourself into thinking that they can replace it.

Is it harder to develop a site-specific Emergency Action Plan, to conduct your own training and drills? You bet it is.

Is it worth it? Perhaps that depends upon how much you're willing to gamble.

Inspection of Exits

This Standard does not mention exits, nor does it reference OSHA Title 29 CFR Part 1910.37, Means of Egress. This is odd because exits are absolutely essential to evacuation.

An evacuation won't go very well if aisleways or exits are blocked by machinery, pallets, or boxes. Employees are likely to panic if they come to an exit door only to find it locked, or if they go to an unmarked door, thinking it is an exit, only to find it leads to a storage closet or boiler room.

If you are unfamiliar with 1910.37 and neither OSHA nor the local fire department has paid you a visit in recent history, the condition of your exits may not be up to par. How do you find out if there is anything you should do to improve them? Take a walk around your building and look at them. A good time to do this is when you are drawing up the evacuation route maps that are intended to lead employees to open, accessible exits.

Ask yourself the following questions as you tour your building:

Is there a clear aisleway to each exit?
Are the aisleways marked?
Does equipment ever block the exit?
Are all exits to the outside marked with an illuminated sign?
Are any exits to the outside ever locked?
If certain exits are locked during certain shifts, are employees on those shifts given evacuation routes to different exits?

If employees must move through internal room exits to get to an exit to the outside, are those internal exit doors labeled "EXIT," and are all other doors in those internal rooms labeled according to whatever is on the other side; e.g. "CLOSET," "BREAKROOM" ?

The purpose of these questions is to prompt you to examine your workplace so as to ensure that every employee on every shift has a clear path from his or her work station to an unlocked exit to the outside of the building at all times. You may tour your building today and find that this is true. You may find that it isn't true today, but it's usually true. Or you may discover that it has never been true. Regardless of the present state of your exits and aisleways, the only way to insure that they are clear and open every day your company operates is to conduct regular inspections.

Case in Point: A southeastern food processing plant employs 225 people at a sprawling rural facility. Most of the workers make minimum wage. The 20-year-old plant has not been inspected by any agency in nearly 12 years. It is equipped with the original sprinkler system, which does not have an automatic heat detector. There are two fire extinguishers in the entire plant—one in the lunch room and one in the front office. The company does not have an emergency evacuation plan, nor does anyone at the plant inspect exits to make sure they are open and accessible. When a fire starts at the facility on a hot summer afternoon, 25 people are killed trying to escape through two emergency exits. One is locked. The other is blocked.

FIRE PREVENTION PLANS

Regulatory Code:

> 1910.38 (b) Fire Prevention Plan (1) Scope and Application. The Fire Prevention Plan shall be in writing (except for those employers with 10 or fewer employees who may communicate the requirements of the Plan orally to their employees). (2) Elements. The following elements, at a minimum, shall be included in the Fire Prevention Plan: (i) A list of the major workplace fire hazards and their proper handling and storage procedures, potential ignition sources (such as welding, smoking and others) and their control procedures, and the type of fire protection equipment or systems which can control a fire involving them; (ii) Names or regular job titles of those personnel responsible for maintenance of equipment and systems installed to prevent or control ignitions or fires; and (iii) Names or regular job titles of those personnel responsible for control of fuel source hazards. (3) Housekeeping. The employer shall control accumulations of flammable and combustible waste materials and residues so that they do not contribute to a fire emergency. The housekeeping procedures shall be included in the written Plan. (4) Training. (i) The employer shall appraise employees of the fire hazards of the materials and processes to which they are exposed. (ii) The employer shall review with each employee upon initial assignment those parts of the Fire Prevention

Plan which the employee must know to protect the employee in the event of an emergency. The written Plan shall be kept in the workplace and made available for employee review. (5) Maintenance. The employer shall regularly and properly maintain, according to established procedures, equipment and systems installed in heat producing equipment to prevent accidental ignition of combustible materials. The maintenance procedures shall be included in the written Fire Prevention Plan.

English: What is the purpose of the Fire Prevention Plan? To prevent fire, of course. But how? By *identifying* workplace fire hazards and training your employees to recognize and control them through *maintenance, inspection,* and *housekeeping* procedures.

Developing a Plan
The key to your Fire Prevention Plan is identifying and listing workplace fire hazards. Let's return to Department 31 at A-1 Manufacturing, introduced on page 97. If you were responsible for developing a Fire Prevention Plan for that company, what would you include on a list of workplace fire hazards?

1. Solvents in paint kitchen.
2. Hazardous waste outside paint kitchen.
3. Propane tank on forklift.

One section of your Plan would answer the question: What are the proper handling and storage procedures for these materials? For solvents and hazardous wastes, lids and bungs should be kept in place when not in use. Containers should be monitored for leaks and spills and should be cleaned up immediately when they are discovered. For the propane tank, valves and hoses should be inspected and monitored for leaks.

For each material on your list, you also would answer these questions:

What are potential ignition sources? Cigarettes, sparks, forklift backfire.
What procedures should be followed to control ignition? Smoking should not be allowed in Department 31. Explosion proof pumps should be used in paint kitchen. The forklift should be tuned-up routinely.
What type of fire control systems could put out a fire? A carbon dioxide fire extinguisher or heat-sensitive fire blanket.

This is one format for dealing with the list of fire hazards required in paragraph (2)(i). Remember, the list is intended to be used as a tool for training your employees, so pay special attention to *procedures* for avoiding fires. Be as specific as possible. For example, if you have an area where workers are supposed to put lids on containers of flammable liquids when they are finished with them, put them back in a flammable storage cabinet, and then close the door of the cabinet, include the *whole* procedure as part of proper handling and storage of flammable liquids.

Chemicals may not be the only fire hazards you need to list. Other fire hazards might include solvent-soaked wipes or rags, discarded paint filters, floor or overhead space heaters, welding torches, pilot lights, heating elements on process equipment, extension cords, exposed electrical wiring, sawdust, and just plain old combustible material, such as paper, cardboard, wood, or regular trash. All of these things must be stored, handled, operated, and controlled in a certain way in order to avoid fire.

As with the Emergency Action Plan, the Fire Prevention Plan will benefit from the knowledge and experience of your employees. Get supervisors and others involved in the process of hazard identification. Ask them to tell you the best way to handle certain materials so as to avoid fire. Ask them to identify sources of ignition. Have them tell you how those sources can be controlled. As part of their training, you may want to ask each department to come up with a list of fire hazards, procedures, and controls, or conversely, you might develop the list and have them critique it.

There are many ways to approach fire prevention with your personnel. Awareness of fire hazards comes up in at least two other regulations of interest to most employers. Identifying hazardous materials that are fire hazards and talking about proper storage and handling of those materials is part of Hazard Communication training. Identifying procedures for storing and handling hazardous wastes that are fire hazards is part of the Personnel Training requirements of RCRA. You might use training under these regulations as an opportunity to lay the groundwork for or identify what items need to be included in your Fire Prevention Plan.

The Role of Maintenance and Inspection
The rest of the Standard focuses on maintenance and inspection as keys to preventing and controlling fires. Procedures for each must be included in your Plan. The Standard identifies two areas of your operation where maintenance and inspection procedures are crucial: (1) Fire prevention or control equipment; and (2) Fuel sources.

Fire Prevention or Control Equipment
First, what is fire prevention equipment? Usually, it is equipment that is installed on heat-producing machinery to *guard against fire*. It could be a heat monitor or detector, a cooling system, a physical barrier, or a heat or smoke alarm. Fire control equipment is what you use to stop a fire *after it has started*. Such equipment might include fire extinguishers, sprinklers, and fire hoses. The Standard mandates that the maintenance of fire prevention and control equipment is the dual responsibility of both employer and employee.

Employer—It is the employer's ultimate responsibility to maintain this equipment so that it works. In order to do this, the Standard says that the employer must develop written standard operating procedures for the maintenance of this equipment. He or she must also assign specific employees with the job of carrying out the maintenance procedures and train them in how to do it.

Employee—It is the responsibility of whoever has been given the job of maintaining fire prevention and control equipment to follow the standard operating procedures in the Fire Prevention Plan.

The most important element of maintenance procedures is an inspection schedule. Although the Standard does not use the word "inspection," it is impossible to run an effective maintenance program without thorough, routine inspections. Why? Well, the term "maintain" means to keep in good working order. In order to keep equipment in good working order you need to, quite literally, fix it before it breaks. Maintenance personnel are not supposed to be crisis managers. The purpose of having maintenance personnel is not to have somebody on hand to fix equipment when it breaks down, but, ideally, to have somebody whose sole job is to prevent equipment from breaking down in the first place. In order to do this, they can't sit in an office waiting for repair orders; they have to be out in the workplace inspecting equipment.

As for fire control equipment, since it is in your facility for the sole purpose of being used in a fire emergency, it is crucial that it is in working order at all times. Broken or malfunctioning emergency equipment is a serious liability and can mean the difference between a small, easily controlled blaze and a disastrous inferno. This equipment must be routinely inspected to insure that it has been recharged since its last use and that it is not defective.

Written maintenance and inspection procedures that designate specific personnel to perform them should carry the weight of policy at your company, especially if they were arrived at through a consensus process. As with the Emergency Action Plan, this consensus can be arrived at through training, both under this Standard and RCRA. Part 264.16(3) of RCRA's Personnel Training requires that employees are familiar with the procedures for using, inspecting, repairing, and replacing facility emergency and monitoring equipment.

If, as part of your RCRA compliance, you already have a maintenance and inspection program for equipment involving hazardous waste, then that program can be used as the foundation of a plant-wide program for fire prevention under this OSHA Standard. Or, working in reverse, as you develop a maintenance program for fire prevention and control equipment under this Standard, don't leave out equipment that involves hazardous waste. Include it, but earmark it as part of your RCRA Contingency Plan and Personnel Training.

Fuel Sources
What are fuel sources? Remember the fire triangle? Three things must be present to support fire: oxygen, heat, and fuel. Fuel sources can be a wide variety of things

at your workplace: flammable and combustible liquids, compressed gases, accumulations of combustible trash and dust. Again, control of fuel source hazards is the dual responsibility of the employer and the employee.

Employer: Ultimately, the responsibility falls upon the employer to operate his or her facility in such a way that fuel source hazards are controlled. The Standard requires the employer to develop written housekeeping procedures to control accumulations of flammable and combustible materials that might cause or add to a fire emergency. The employer also is required to designate personnel who will be responsible for making sure that fuel sources are controlled and to train them in how to do it.

Employee: It is the job of whoever has been designated to control fuel source hazards to maintain the facility in accordance with the housekeeping procedures included in the Fire Prevention Plan.

As with maintenance of fire prevention and control equipment, the maintenance of your facility in accordance with an established "standard of housekeeping" requires routine inspection and follow through. In other words, your housekeeping procedures must be more than just a generalized policy statement, such as "We will operate Smith Printing Company in a manner that will prevent workplace fire through the control of accumulations of flammable and combustible waste materials near ignition sources." This is the kind of bureaucratic non-statement that encourages cynicism in all who hear it. It makes you want to say, "Yeah, right. More regulatory mumbo jumbo, huh?"

In developing housekeeping procedures for fire prevention, look back at your list of workplace fire hazards and ignition sources. Or better yet, *as you develop the list* with your personnel, ask them how housekeeping procedures might be used to monitor or reduce fire hazards. The procedures should be specific enough to insure their effectiveness. (See Chapter 7 for more on housekeeping procedures.)

For instance, you might develop a procedure that no more than two cans of paint or thinner are allowed to be taken out of a flammable storage cabinet at one time. You might develop a procedure that "empty" chemical containers are not allowed to accumulate in an area where welding is conducted, or that combustible trash cannot be "thrown away" in 55-gallon drums still labeled as containing flammable solvent; only drums clearly labeled "Trash" can be used as "trash cans."

Key to successful fire prevention is the idea that all employees must be familiar with the housekeeping procedures that apply to them. Part of their fire prevention training, which overlaps with both Hazard Communication and RCRA Personnel Training, must include *how to do their job so as to prevent fire.* The Hazard Communication Standard requires that employees understand the physical hazards of materials—whether or not they are flammable, combustible, explosive, or reactive—so that they can work in such a manner that will avoid fire, explosion, or reaction. Inherent in RCRA's requirements is the mandate that employers maintain and operate their facilities so as to limit the possibility of fire or other emergency. One major way in which this can be accomplished is through employee knowledge of emergency prevention or, in simpler terms, housekeeping procedures.

The job of the special personnel designated to control fuel source hazards is to inspect the facility to insure that these housekeeping procedures are followed. Whenever procedures aren't being followed, it is also their job to remedy the situation. If, during a housekeeping inspection, four or five containers of flammable liquids are found sitting out on a counter where there are supposed to be only two, this needs to be brought to the attention of someone who can fix it. Depending on the inspector's position of authority within the company, he or she may bring it to the attention of the Supervisor, the Safety Manager, or the employee who works at that station.

Again, the purpose of housekeeping procedures and inspections is to maintain the facility in such a condition that fire hazards are minimized. If you or your personnel are documenting housekeeping deficiencies in your inspections and these deficiencies remain unremedied, obviously your procedures are not achieving their intended result.

As with the Emergency Action Plan, including your personnel in the development of the Fire Prevention Plan is the most effective way to train them. *Everyone* employed at your company needs to be able to recognize fire hazards, ignition sources, and fire control equipment. Each individual also needs to know the specific housekeeping procedures to be followed to minimize fire risk. Through training, the Fire Prevention Plan should bring together elements of your Hazard Communication and RCRA Personnel Training. Through training, your personnel should come to see fire prevention not as a collection of activities separate from the management of hazardous materials and hazardous wastes, but as an approach toward work that cannot be separated from any aspect of it.

6

More Than Maintenance

*Nothing is more terrible
than activity without insight.*

Thomas Carlyle

OSHA's Standard governing the Control of Hazardous Energy or Lockout/Tagout, as it is popularly called, is a formalized system for organizing and controlling maintenance operations to insure that they are performed safely. Lockout/Tagout is to hazardous energy what the Hazard Communication Standard is to hazardous materials. In fact, Lockout/Tagout could very well be called Right-to-Know for hazardous energy in the workplace.

An often misunderstood regulation, Lockout/Tagout is intended as a guide for employers to use in developing procedures for avoiding the *unexpected* energization (or connection to an energy source) of equipment while it is being serviced or repaired. *Unexpected* is the key word here. The primary goal of the Standard is to aid employers in developing procedures that will prevent equipment or machines from being unexpectedly turned on, started, opened, or released. If there is no conceivable way for start-up or energization of the equipment to occur unexpectedly (without the knowledge of the person conducting the maintenance or repairs), then lockout or tagout procedures are not required. It is important to remember, however, that even in cases where the energization of equipment is under the complete control of the maintenance technician or equipment operator, standard operating procedures for notifying other employees of equipment start-up and for using machine guards and safety devices still need to be established.

On the surface, this Standard has little to do with Hazard Communication or Emergency and Fire Prevention Plans. However, on closer reading of the definitions at the beginning of the regulation, it is clear that there is a direct relationship.

First, although electrical energy is the first energy source that comes to mind when you think about energy hazards in the workplace, the Standard defines energy source much more broadly: *any source of electrical, mechanical, hydraulic, pneumatic, chemical, thermal, or other energy.*

Second, the Standard defines servicing and maintenance as: *workplace activities, such as constructing, installing, setting up, adjusting, inspecting, modifying, and maintaining and/or servicing machines or equipment where the employee may be exposed to the unexpected energization or start-up of the equipment or release of hazardous energy.*

When does Lockout/Tagout overlap with Hazard Communication?

1. If, during maintenance on a certain piece of equipment, an employee is required to shut off and lock or tag out a valve or other device that controls the flow of chemicals into the area where he or she is working.
2. If there is a potential for release of stored chemical energy (such as release of gas or pressure due to heat build-up or chemical reaction) in the equipment he or she is repairing.
3. If, during maintenance on a certain piece of equipment requiring lockout or tagout, he or she is required to utilize hazardous materials or activate systems involving hazardous materials.

In all of these cases, the employee conducting the servicing needs to know how to isolate both him or herself and the equipment on which he or she is working from potential exposure to chemical energy. Training in chemical hazards will help employees take the proper precautions, both by locking or tagging chemical source controls and by wearing adequate protective equipment.

When does Lockout/Tagout overlap with Emergency and Fire Prevention Plans?

1. The definition for servicing equipment includes inspecting, adjusting, modifying, and maintaining it. Whenever maintenance and inspection of fire prevention and control equipment is required, there is a potential for overlap with the Lockout/Tagout Rule. For example, employees required, under the company Fire Prevention Plan, to conduct inspections or repairs on temperature limit devices on high-temperature gluing equipment or on flame failure and flashback arrester devices on furnaces need to be trained in the proper lockout or tagout procedures to follow during these inspections and repairs, to insure that they will not risk exposure to an unexpected release of hazardous energy.
2. Also included in the definition of servicing and/or maintenance is the setting up of equipment for normal production operations. When machines and equipment must be must be reenergized after actual emergencies or emergency drills, there is potential for overlap with Lockout/Tagout. For example, employees required, under the company Emergency Action Plan, to shut down operations when evacuation is required must be trained in safe procedures for setting up and reenergizing equipment, including inspection of the workarea and verification of the safety of other potentially exposed employees.

THE LOCKOUT/TAGOUT RULE

What does the Control of Hazardous Energy, OSHA's Lockout/Tagout Rule, require?

Although the language of this regulation is sometimes highly technical and often seems esoteric, or remote from the hands and knees, grease and grime rush of getting a piece of equipment up and running as fast as possible, it is important to keep in mind that the purpose of this regulation is *not* to tie up your maintenance personnel in arduous procedures and red tape. The purpose, simply, is to prevent one of them from becoming a statistic on the annual list of injuries and deaths that occur due to the unexpected energization of equipment. And, like every other Standard discussed in this book, the only way to insure that your program fulfills its intended purpose is to make it practical, understandable, and do-able—not arduous at all.

Regulatory Code:

1910.147 The Control of Hazardous Energy (Lockout/Tagout). (a) Scope, Application and Purpose. (1) Scope. (i) This Standard covers the servicing and maintenance of machines and equipment in which the unexpected energization or start up of the machines or equipment, or release of stored energy could cause injury to employees. This Standard establishes minimum performance requirements for the control of such hazardous energy. (2) Application. (i) This Standard applies to the control of energy during servicing and/or maintenance of machines and equipment. (ii) Normal production operations are not covered by this Standard. Servicing and/or maintenance which takes place during normal production operations is covered by this Standard only if: (A) An employee is required to remove or bypass a guard or other safety device; (B) An employee is required to place any part of his or her body into an area on a machine or piece of equipment where work is actually performed upon the material being processed (point of operation) or where an associated danger zone exists during a machine operating cycle. Note: Minor tool changes and adjustments and other minor servicing activities which take place during normal production operations, are not covered by this Standard if they are routine, repetitive and integral to the use of the equipment for production, provided that the work is performed using alternative measures which provide effective protection. (iii) This Standard does not apply to the following: (A) Work on cord and plug connected electric equipment for which exposure to the hazards of unexpected energization or start up of the equipment is controlled by the unplugging of the equipment from the energy source and the plug being under the exclusive control of the employee performing the servicing or maintenance. (B) Hot tap operations involving transmission and distribution systems for substances such as gas, steam, water, or petroleum products when they are performed on pressurized pipelines, provided that the employer demonstrates that (1) continuity of service is essential; (2) shutdown of the system is impractical; and (3) documented procedures are followed, and special equipment is used which will provide proven effective protection for employees. (3) Purpose. (i) This section requires employers to establish a program and utilize procedures for affixing appropriate lockout devices or tagout devices to energy isolating devices and to otherwise disable machines or equipment to prevent unexpected energization, start-up or release of stored energy in order to prevent injury to employees. (ii) When other Standards in this part require the use of lockout or tagout, they shall be used and supplemented by the procedural and training requirements of this section.

English: The first question on everyone's mind is: *Do I have to comply with this Standard?* In order to figure that out, let's begin by identifying who *doesn't* have to comply.

What's *Not* Covered

In paragraph (a)(1)(ii) the Standard says that it does not cover construction, agriculture, and maritime employment; electric utilities, including installations for power generation, transmission and distribution, and related equipment for communication or metering; electrical hazards from work on, near, or with conductors or equipment in electric utilities; or oil and gas well drilling and servicing. If your company does not fall into one of these groups, then you are covered by this Standard.

Now, the next question is: *When am I covered*? In order to figure that out, you have to pull out the sections of the Standard that identify what's *not covered.*

Paragraph (a)(2)(ii) says that *normal production operations* are not covered by this Standard as long as the requirements of Title 29 CFR Subpart O, Machinery and Machine Guarding, are met. Further, servicing and maintenance during normal operations that does not require an employee to remove a guard, bypass a safety device, or put his or her body on a point of operation or into a danger zone is not covered by the Standard. In other words, as long as the maintenance activity is routine and integral to the use of the equipment for production, such as a tool change, a minor adjustment or lubrication, and adequate guarding and procedural protection is provided (in accordance with Subpart O), then it is not covered by Lockout/Tagout. The note about Subpart O is the kicker. If you didn't have to meet machine guarding requirements, then you could, in effect, label all maintenance activities "normal" or "routine" and call yourself exempt from Lockout/Tagout.

Ah, but OSHA won't let you take the easy route. Paragraph (a)(2) makes it clear that during *all servicing activities,* whether normal and routine or unusual and infrequent, employees must be adequately protected from hazardous energy, either through Subpart O in the case of the routine activity, or through Lockout/Tagout in the case of infrequent or non-routine maintenance.

There are two other times that the Standard *may not* apply: when work is being done on cord and plug connected electric equipment and when hot tap procedures are conducted. When any type of servicing, whether routine or not, is conducted on cord and plug connected equipment, such as sewing machines, staplers, rewinders, cameras and film processing equipment, drills, copiers, and small presses, and energy is isolated from the equipment by simply unplugging it, then Lockout/Tagout does not apply *as long as* the plug is *under the control of the person conducting the service.* In other words, if a piece of equipment has been unplugged for servicing and the plug is in another room or beyond the sight of the person doing the servicing, then there is a chance that someone may walk by and decide to plug in the cord, *especially if it is usually plugged in.* The person walking by is likely to think that he or she is doing the equipment operator a favor by plugging in the cord. If such a situation exists, then Lockout/Tagout procedures need to be applied. If, however, the cord remains under the control of the person doing the servicing, then Lockout/Tagout is unnecessary.

A hot tap is a procedure that involves welding on a pipeline, vessel, or tank that is under pressure. Hot tap operations on pressurized pipelines distributing gas, steam, water, or petroleum products are exempt from coverage under Lockout/Tagout as long as the employer can demonstrate that continuity of service is essential, shutdown of the distribution system is impractical, and documented safety procedures are followed to protect personnel from unexpected energization of the equipment being used.

Unless every piece of equipment at your company is cord and plug connected and your repair personnel keep the plugs in their possession while conducting repairs or the only activities you conduct that involve energy are hot tap operations, then you are, at some level, covered by this Standard.

Now that you know where you stand, let's answer the final question from this section: *What is Lockout/Tagout supposed to accomplish?*

It is supposed to help employers prevent injury to employees by establishing procedures for isolating machines and equipment from their energy sources to insure that they will not be *unexpectedly energized* (started, turned on, released, operated, opened) while maintenance or servicing is conducted. The procedures developed must be workplace-specific, appropriate to the servicing and maintenance activities conducted and the hazards encountered at your particular workplace. Bottom line, compliance with Lockout/Tagout is supposed to make your workplace a safer place. If your compliance program doesn't accomplish this, then you need to stop and reevaluate what you are doing with this specific goal in mind.

Regulatory Code:

> 1910.147(c) General—(1) Energy Control Program. The employer shall establish a program consisting of energy control procedures and employee training to ensure that before any employee performs any servicing or maintenance on a machine or equipment where the unexpected energizing, start-up or release of stored energy could occur and cause injury, the machine or equipment shall be isolated and rendered inoperative in accordance with paragraph (c)(4) of this section. (2) Lockout/Tagout. (i) If an energy isolating device is not capable of being locked out, the employer's Energy Control Pro-

gram under paragraph (c)(1) of this section shall utilize a tagout system. (ii) If an energy isolating device is capable of being locked out, the employer's Energy Control Program under paragraph (c)(1) of this section shall utilize lockout, unless the employer can demonstrate that the utilization of a tagout system will provide full employee protection as set forth in paragraph (c)(3) of this section. (iii) Whenever major replacement, repair, renovation or modification of machines or equipment is performed and whenever new machines or equipment are installed, energy isolating devices for such machines or equipment shall be designed to accept a lockout device. (3) Full Employee Protection. (i) When a tagout device is used on an energy isolating device which is capable of being locked out, the tagout device shall be attached at the same location that the lockout device would have been attached and the employer shall demonstrate that the tagout program will provide a level of safety equivalent to that obtained by using a lockout program. (ii) In demonstrating that a level of safety is achieved in the tagout program which is equivalent to the level of safety obtained by using a lockout program, the employer shall demonstrate full compliance with all tagout related provisions of this standard together with such additional elements as are necessary to provide the equivalent safety available from the use of a lockout device. Additional means to be considered as part of the demonstration of full employee protection shall include the implementation of additional safety measures such as the removal of an isolating circuit element, blocking of a controlling switch, opening of an extra disconnecting valve or the removal of a valve handle to reduce the likelihood of inadvertent energization.

English: This section serves as a preamble to the heart of the Standard, which is the written energy control procedure and employee training.

Make It Policy

The term "program" used throughout this section can be substituted with "policy." Read this way, it becomes clear that what the Standard is asking you to do is to establish three Energy Control Policies at your company:

1. Before any employee performs any servicing on a machine or equipment where the unexpected energizing, start-up, or release of stored energy could occur and cause injury, the machine or equipment will be isolated by lockout or tagout.
2. In order to provide full employee protection, lockout will be used as the preferred method of energy isolation.
3. When lockout is impossible, tagout will be used in such a manner that it will provide the same level of protection achieved by lockout.

This section of the Standard is asking you to adopt a mindset or attitude toward the energy control procedures you will develop that will insure that they provide *full employee protection.* It is almost as if this section is saying: *As you look around your workplace and identify the equipment and procedures covered by this Standard, keep in mind that the greatest protection from unexpected start-up of equip-*

ment is provided when equipment is physically isolated from its energy source. Wherever possible, you need to adopt a lockout policy. If you can't lockout, carefully examine the equipment and try to figure out ways to enhance the effectiveness of tags and labels so that, while the equipment is not locked in a disconnected position, it is made quite difficult if not impossible to inadvertently turn on.

If, when you begin the development of your step-by-step energy procedures, you already have a full employee protection policy firm in your mind, writing the procedures will be easier (because they will be focused toward one goal), and the procedures, once written, will be more effective in preventing unexpected energization because they will be enforceable as *policy.*

Regulatory Code:

1910.147 (4) Energy Control Procedure. (i) Procedures shall be developed, documented and utilized for the control of potentially hazardous energy when employees are engaged in the activities covered by this section. (ii) The procedures shall clearly and specifically outline the scope, purpose, authorization, rules, and techniques to be utilized for the control of hazardous energy, and the means to enforce compliance including, but not limited to, the following: (A) A specific statement of the intended use of the procedure; (B) Specific procedural steps for shutting down, isolating, blocking and securing machines or equipment to control hazardous energy; (C) Specific procedural steps for the placement, remodel and transfer of lockout devices or tagout devices and the responsibility for them; (D) Specific requirements for testing a machine or equipment to determine and verify the effectiveness of lockout devices, tagout devices and other energy control measures. (*Note: Skip to paragraph (d) because it specifies the steps that must be included in your procedure.*) (d) Application of Control. The established procedure for the application of energy control (implementation of lockout or tagout system procedures) shall cover the following elements and actions and shall be done in the following sequence: (1) Preparation for shutdown. Before an authorized or affected employee turns off a machine or equipment, the authorized employee shall have knowledge of the type and magnitude of the energy, the hazards of the energy to be controlled and the method or means to control the energy. (2) Machine or equipment shutdown. The machine or equipment shall be turned off or shut down using the procedures required by this Standard. An orderly shutdown must be utilized to avoid any additional or increased hazard(s) to employees as a result of deenergization. (3) Machine or equipment isolation. All energy isolating devices that are needed to control the energy to the machine or equipment shall be physically located and operated in such a manner as to isolate the machine or equipment from the energy source(s). (4) Lockout or tagout device application. (i) Lockout or tagout devices shall be affixed to each energy isolating device by authorized employees. (ii) Lockout devices, where used, shall be affixed in a manner that will hold the energy isolating devices in a "safe" or "off" position. (iii) Tagout devices, where used, shall be affixed in such a manner as will clearly indicate that the operation or movement of energy isolating devices from the "safe" or "off" position is prohibited. (A) Where tagout devices are used with energy isolating devices designed with the capability of being locked, the tag attachment shall be fastened at the same point at which the lock would have been attached. (B) Where a tag cannot be affixed directly to the energy isolating device, the tag shall be located as close as safely possible to the device, in a position that will be immediately obvious to anyone attempting to operate the device. (5) Stored Energy. (i) Following the application of lockout or tagout devices to energy isolating devices, all potentially hazardous stored or residual energy shall be relieved, disconnected, restrained and otherwise rendered safe. (ii) If there is a possibility of reaccumulation of stored energy to a hazardous level, verification of isolation shall be continued until the servicing or maintenance is completed, or until the possibility of such accumulation no longer exists. (6) Verification of isolation. Prior to starting work on machines or equipment that have been locked or tagged out, the authorized employee shall verify that isolation and deenergization of the machine or equipment have been accomplished.

English: As presented in the Standard, the development of an energy control procedure appears overwhelmingly complicated. In order to cut through the jargon and end up with a procedure that your employees can use, you have to take it one small step at a time (see Figure 6-1).

Developing a Procedure

Identify Authorized and Affected Employees
The Standard defines an "authorized" employee as a person who has been given the authority to use a lockout or tagout procedure when performing maintenance on machinery or equipment. The authorized employee may be a maintenance employee or a contract employee whose sole job is to service and repair machines and equipment, or may be the equipment operator if the job description includes maintenance activities that require lockout or tagout.

"Affected" employees are all other employees who work where service or maintenance of equipment is conducted under lockout or tagout and are thus affected by lockout or tagout procedures and restrictions. Your energy control procedure should include a listing of the names and job titles of authorized and affected employees.

Identify Energy Control Hazards
Although the Standard does not specifically require it, it is unreasonable to attempt to write a procedure for the control of hazardous energy at your workplace before you have identified both the *source(s) of the hazard(s)* and *when your employees may be exposed to them.* How do you do this? First identify the machines and equipment that are connected to an energy source. Then identify the specific servicing and maintenance activities that will require lockout or tagout in order to provide full employee protection under your Energy Control Policy. One word of advice: It is virtually impossible to do this without the input of or the personnel who regularly operate the equipment (affected employees), as well as that of

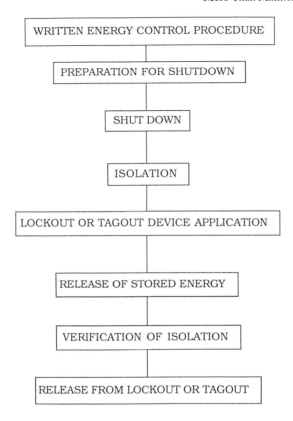

Figure 6-1. Flowchart of Elements of Energy Control Procedure.

anyone who will conduct non-routine servicing and maintenance under lockout or tagout (authorized employees).

A convenient way to identify the hazards of normal operations is to examine each machine individually or by department and ask the personnel who normally operate the equipment to describe the routine maintenance and servicing activities (such as lubrication or tool changes) that they conduct. Have them describe the steps they follow to conduct the routine maintenance. Do they remove any guards? Do they place their hands in a danger zone? If they work on cord and plug connected equipment, do they unplug the equipment before they conduct the routine service; and do they keep the plug within their control?

If you determine that your personnel are adequately protected during these routine operations, then, as mentioned earlier, lockout or tagout is unnecessary. However, it is still very important to identify in your energy control procedures what these normal activities are and what safety procedures must be followed in lieu of lockout or tagout. If the routine maintenance procedures your personnel are following are revealed to be *unsafe* according to this Standard and Subpart 0, then you have two choices: Either alter the routine maintenance procedures so that full protection is provided without lockout or tagout, or require those personnel to use lockout or tagout during routine maintenance.

Sample program excerpt: "Normal, routine operator maintenance conducted by press operators in Department 1 consists of replacing rollers, lubricating rollers and removing paper jams. In all of these cases the emergency stop button and run-jog-stop switch is under the immediate control of the operator so that unexpected or inadvertent start-up of the press is unlikely, except in the case of the operator him or herself making an error. In order to protect against such error, the removal of machine guarding devices is not allowed during routine maintenance and operators have been instructed to avoid placing their fingers and hands in a danger zone (such as against a moving roller). As long as these requirements are met, then lockout or tagout is not required. However, whenever these safety requirements cannot be met, then lockout or tagout procedures as described for non-routine maintenance must be followed in order to insure that the operator is fully protected."

For non-routine maintenance, or any servicing activity that requires the employee to remove a guard or safety device, place a body part into a point of operation or a danger zone, or work in an area where the controls for the machine or equipment are out of reach or sight, it is still an excellent idea to have the personnel who conduct the servicing describe *exactly what they do*. Again, take the machines and equipment at your company individually or by department and compile a list of maintenance and servicing activities that will require either lockout or tagout. Include these non-routine maintenance and servicing activities as part of the hazard identification section of your energy control procedure.

Sample program excerpt: "Non-routine maintenance of presses in Department 1, such as the repair or replacement of motors or electrical elements, requires the use of lockout procedures by the authorized employee conducting the maintenance."

Identify Energy Isolating Devices

Remember, in order to isolate a machine from its energy source, you have to do more than turn it off at its primary operating controls; you have to block energy from entering the machine at all. In other words, you can deactivate a copy machine by turning it off. But in order to isolate it from its energy source, it must be unplugged. You can deactivate a lathe by pressing the stop button. But in order to isolate it from its energy source, you must place the circuit breaker in the "off" position.

In order to specify the procedural steps that your personnel must follow to shut down, isolate, block, or secure the various machines or equipment at your company, you have to figure out how energy can be isolated from each piece of machinery or equipment for which lockout or tagout may be required.

More Than Maintenance   119

A convenient way to do this is to set up a chart that lists each piece of equipment connected to an energy source in one column and that has three columns with the headings First, Second, and Third Energy Isolating Device (see Figure 6-2). Much more effectively than a narrative format, this chart indicates the methods available to an employee for shutting down a press and isolating it from its energy source.

The isolating devices will vary depending on the energy sources that are present in any given piece of equipment. For process equipment involving possible chemical exposure, such as a vacuum tank or hazardous waste handling equipment, the energy isolating mechanisms would include not only circuit breakers and main cut-offs, but valves and pumps controlling chemical input. For equipment where fire is present, such as a gas heater or kiln, energy isolating devices might include valves on fuel lines or even the physical removal of the fuel source from the equipment.

Presenting energy isolation alternatives in this format provides you with an excellent tool for classroom training. If you have a maintenance staff, you might decide to provide them with a notebook of these charts and to add them to other standard operating procedures for maintenance, inspection, and emergency response. Additionally, if you use outside contractors to conduct your non-routine maintenance and servicing, you can use these charts as part of the program you provide to inform them of the lockout or tagout procedures you expect them to follow. The topic of providing information to contractors is discussed in greater detail at the end of this chapter.

In identifying energy isolating devices there is one problem so obvious that it is often overlooked: *You can't identify what you can't find.* Many an enthusiastic Lockout/Tagout program has been held up for weeks, even months, because, when

MID-SOUTH PRINTING COMPANY
ENERGY ISOLATING DEVICES CHART

EQUIPMENT	1st DEVICE	2nd DEVICE	3rd DEVICE
Step & Repeat	Master power control on monitor. Red Stop buttons on four corners.	Switch #3, 5, 7 in Box 11 by Prep. Dept.	Front Building Main (Box 1).
MOF Heidelberg (5-color)	On/Off switch. Red Safety stop switches. Red master power cut-off on controller panel.	Main switch in Box #23 by Green Cutter.	Back Building Main (Box 30).
SORKZ Heidelberg (2-color)	On/Off switch. Red Safety stop swtiches. Red master power cut-off on controller panel.	Switch #19, 21, 23 in Box 8 by Prep. Dept.	Back Building Main (Box 30).
SORKZ Heidelberg (1-color)	On/Off switch. Red Safety stop switches. Red master power cut-off on controller panel.	Switch #8, 10, 12 in Box 4B by Prep. Dept.	Back Building Main (Box 30).

Figure 6-2.   Energy Isolating Device Chart.

they got right down to it, no one at the company, from the President on down the line, knew where to go to isolate equipment from its energy source.

Case in Point: At a mid-sized commercial printing company, the Production Manager has hired a consultant to develop a Lockout/Tagout Program.

On a tour of the plant for the purpose of identifying energy isolating devices for presses, cutters, and bindery machines, the Production Manager is confident that he knows which circuit breakers correspond with which machines. Although few breakers are labeled, the Production Manager seems to know how everything is hooked up by heart. The consultant writes it all down and beings making charts and diagrams.

Everything would have gone along like this except for the fact that when the consultant has a question and can't get in touch with the Production Manager, she talks with the Maintenance Supervisor. During their discussion, it becomes evident that most of the information given to the consultant by the Production Manager is wrong. The program is back to square one. As a result of 30 years of changes, building expansions, and the rearrangement of the departments, it seems that no one at the company knows how anything is energized. Nothing is labeled correctly.

Ultimately, the consultant must make four separate trips to the company and spend a total of ten hours with the Maintenance Supervisor. Wreaking havoc on production schedules, they end up having to turn on each machine individually and then flip switches in circuit breaker boxes until they find the one that turns it off. Adding to the confusion, it is discovered that in nearly every case the electrical panels in any given department correspond to equipment not in that department but someplace else.

Having unlabeled circuit breakers, transformers, electrical mains, fuel supplies, inlet valves, and pipes (which are required to be labeled under the Hazard Communication Standard, see Chapter 3) is a dangerous position to be in, both from the standpoint of controlling hazardous energy and preventing emergencies and fires. Nevertheless, it is something that employers have been known to put off, especially if they have never had an emergency. In requiring you to locate and label your energy source controls, this Standard is helping you to meet emergency and fire prevention planning requirements at the same time.

Once you know how to isolate a piece of equipment from its energy source(s), paragraph (d) of the Standard indicates specific procedures you need to include in your written energy control procedure for doing just that. The following three steps apply in all cases where lockout or tagout is required:

1. Preparation for Shutdown—This simply means that, before equipment or machinery is shut down for servicing, the employee turning off the equipment needs to know three things:

a. The type or types of energy that need to be controlled.
b. The hazards of the energy.
c. The methods or devices available for isolating the energy.

Before they turn off equipment for the purpose of conducting maintenance on it, they need to know where all of the energy isolating devices (cut-offs, circuit breakers, valves, etc.) for that equipment are located and how to use them. Where the possibility of exposure to chemicals, fuels, steam, or pressure exists, they need to be trained in the hazards of this energy and how to restrain it.

2. Machine or Equipment Shutdown—This step is included to emphasize the need for planned, orderly equipment shutdown. Although it is of greatest importance in the case of large process equipment with multiple workstations, all affected employees, regardless of equipment size, need to be notified *before* equipment is shut down. If there are any additional hazards that may be caused by the shutdown, such as heat or pressure buildup or the backup of production or waste materials, affected employees need to be aware of these hazards and of any steps they can take to limit them before the shutdown begins. Even if you have an operation where each machine is run by only one employee and each employee will shut down, lockout or tagout, and conduct the maintenance on his or her own equipment, it is still a good idea to have a procedure whereby each employee will notify the other employees in the immediate area of his or her equipment that he or she is planning a shutdown before it actually commences.

3. Machine or Equipment Isolation—Once affected employees have been notified of equipment shutdown and the authorized employee has turned the equipment off through the normal methods, then he or she can locate the appropriate energy isolating devices for the equipment and operate them so that the equipment is disconnected from its energy source.

Identify Lockout and Tagout Procedures

Now that you have identified *when* the equipment at your company will require energy isolation and *how* it can be isolated, it is time to figure out when and how you are going to use lockout and tagout devices to keep that energy in isolation.

First of all, what does "lockout" mean? It means to literally lock an energy isolating device, such as a circuit breaker, in the "off" position or a valve in the "closed" position. This is accomplished when the person conducting the servicing places a lock *to which only he or she has the key or combination* on the energy isolating device in such a way that it cannot be moved.

"Tagout" means placing a tag, label, or sign on equipment or on a specific energy isolating device that warns against energizing, turning on, or releasing the energy in the equipment. Tags might say "Do Not Start," "Do Not Open," "Do Not Close," "Do Not Energize," "Do Not Plug In," "Do Not Operate," or include any other command or warning that is appropriate to your facility and will prevent your employees from operating the machine or equipment. The Standard requires that tags must be substantial enough to prevent accidental removal and to prevent them from becoming illegible in damp or humid conditions or upon exposure to solvents or corrosive chemicals. In other words, taping a hand-scribbled "Out of Order" sign on a piece of equipment isn't good enough. The tag must be durable, all-environment-tolerant and attached to the equipment or isolating device by a self-locking, non-releasable, nylon cable tie. Although many tags on the market include a space for the authorized employee who placed it to sign his or her name, this is not a requirement of the Standard. If you have a large company, you might find it practical to incorporate the signing of tags into your energy control procedures.

Perhaps because the Standard is called Lockout/Tagout, some employers seem to think that both a lock *and* a tag must be used at the same time during maintenance on one piece of equipment. This is not the case. If lockout is used, then the placement of warning tags (labels, signs) on equipment is not required.

Paragraph (c)(5) of the Standard states that it is the employer's responsibility to provide the locking and tagging devices that will work on the equipment at their facility. In other words, if none of your equipment is designed in such a way that it can be locked out, it doesn't do your employees much good if you order a drawer full of locks to "comply" with this Standard. Also, locks and tags should be standardized within your facility so that they are easily identifiable by color, shape, style of print, or format, and they should not be used for anything other than the control of energy during maintenance and servicing. Using them for other purposes only dilutes their power to influence behavior.

As a rule of thumb, if you have identified the need for energy isolation in a certain piece of equipment and that equipment *can be locked,* then it *should be locked.* The reasoning here is that, since tags are essentially warning devices, they do not provide physical restraint. Since it takes approximately the same amount of time to place a lock on an energy isolating device as it takes to place a tag, why not use the lock and enjoy added protection?

The Standard, however, does allow employers to use tags in place of locks whenever they can show that the tags provide the *same level of protection* as locks. The only instance in which this approach seems worth the trouble is when the person conducting the maintenance needs to be able to energize the equipment *during maintenance activities.* In this case, locking the equipment in the off position would be impractical. Such circumstances need to be identified in your energy control procedures along with the precautions service personnel are instructed to take to insure that tags provide adequate protection.

Paragraph (d) of the Standard indicates the procedures an authorized employee needs to take to lockout or tagout equipment after it has been isolated from its energy source(s). However, these procedures are very general. When writing this section of your energy control procedures for the specific equipment at your company, remember that you must indicate when lockout will be used and when tagout will be used. A statement such as "After authorized employees have isolated machines or equipment from their energy source(s), lockout or tagout devices will be affixed to each device in such a manner that will hold the energy isolating device in a "safe" or "off" position" is just a regurgitation of the regulation. It does not reflect what your employees actually are going to do after they shut off circuit breakers or valves.

Like the other regulations covered in this book, the Lockout/Tagout Rule requires you to be site-specific in your approach to compliance. Rather than merely restating the requirements of the Standard in your procedures, you must *apply the requirements* to the equipment and maintenance activities at your company and develop procedures that are appropriate. That's why it is so important to conduct the hazard identification survey mentioned at the beginning of this section.

In order to develop meaningful lockout or tagout procedures, you need to review your chart of equipment and energy isolating devices and determine which equipment can be locked out and *where* it can be locked out. This information can be added to the chart of energy isolating devices in the form of an "L" or the word "lock" placed by each device requiring a lock. If you don't want to use the chart format, you can cover the same information in narrative form.

On the chart in Figure 6-3, you can see that lathe #21 can be locked out at the main circuit breaker on the adjacent wall. Therefore, locking out this machine is the most practical, convenient, and safe method of isolating it from its energy source. Cutter #3, however, does not have an energy isolating device that is capable of being locked out. It will be necessary to tag this machine in order to isolate it. A "T" indicating tag has been placed next to the emergency cutoff for the cutter, shown in the second column.

Where machines such as this cutter must be tagged rather than locked or where machines that can be locked are tagged out of necessity or convenience, the Standard requires you to develop a few additional procedures. For each piece of machinery or equipment to be tagged, you need to indicate *how and where* a tag

**DAVIS & SONS MACHINING COMPANY
ENERGY ISOLATING DEVICES CHART**

EQUIPMENT	1st DEVICE	2nd DEVICE	3rd DEVICE
Lathe #21	On/off switch.	Circuit breaker on adjacent wall (L).	Building main
Lathe #22	On/off switch.	Circuit breaker on SW wall (L).	Building main
5-hole Drill	On/off switch. Safety stop button. Pull plug.	Switch #14 in box on adjacent post (L).	Building main
Cutter #3	On/off switch. Safety stop button.	Emergency power cut-off (T).	Building main

Figure 6-3. Modified Energy Isolating Device Chart.

will be connected to the energy isolating device(s) for the equipment. For instance, in the case of the cutter that cannot be locked out, the tag needs to be placed close enough to the energy isolating device to clearly indicate that it should not be moved from the "off" position. Your procedures need to indicate where the tag should be placed and your authorized personnel need to be trained accordingly. In the case of a printing press that can be locked out, but that is being tagged out because of the nature of the maintenance being conducted, the tag should be attached to the energy isolating device at the same place the lock would be attached. Your written procedures need to reflect this, and, of course, your authorized personnel must be trained to carry it out.

Identify Procedures for Release of Energy and Verification of Isolation

This last step in the development of your energy control procedures may be the most important, because this is where you find out whether or not you have done the other steps correctly.

After an authorized employee has locked or tagged out his or her equipment at the appropriate energy isolating device, the nature of the equipment or process may demand that he or she release or relieve any stored or residual energy that could create a hazard for him or her while working on it. Again, as mentioned earlier, general procedural statements that just restate the regulation do your employees no good and provide you no help in training them. *Specifics* are what matter here.

What is residual or stored energy? It can be a lot of things. Stored energy can exist in springs, elevated machine members, hydraulic systems, rotating flywheels or rollers, or whenever there is the potential for air, gas, steam, or water pressure build-up, chemical reaction, or temperature elevation. Examine each piece of equipment you have on your list. Talk to the people who operate it. Use common sense. If, when the equipment is turned off, it can retain energy or force because an element is left raised in the air, a spring remains retracted, or pressure is left unrelieved, then you need to address how to dissipate or restrain that stored energy in your written procedures. If you don't have any equipment in which stored energy might pose a hazard, then you don't need to address this issue at all.

Procedures for dissipating or restraining stored energy will vary greatly from one machine to the next and from one company to the next. In some cases, such as in dealing with an elevated machine member, there may not be one "best" way for protecting an employee. One company may decide to adopt a procedure where elevated machine parts are blocked from falling by using a structural support. Another company may decide that it is easier to physically reposition the raised element by loosening nuts and bolts. In order to control pressure build-up, you may need a procedure where an authorized employee is required to bleed down a line at certain intervals. Or, if pressure or heat build-up on one system is caused by the operation of another system, then whenever maintenance is conducted on the first system, shutdown of the second system might be made mandatory. The possible scenarios for managing stored energy are virtually endless. The important point to remember is that your written procedures need to include the types of stored energy that may be present in the machinery and equipment *at your company* and the methods that authorized personnel should follow to dissipate or restrain that energy.

Finally, prior to starting to work on machines or equipment that have been locked or tagged out, authorized personnel must follow an established procedure for verifying that the equipment really has been disconnected from its energy source. This means exactly what it says. After disconnecting the energy sources and checking to make sure that no other employees are exposed, authorized personnel need to operate the start button, toggle switch, or other operating controls to make sure that the equipment will not operate. Your written procedures should include an explanation of how energy isolation can be verified for the equipment at your facility. Included in the procedure should be the reminder that, after the test has been conducted, operating controls *must be returned to the neutral or "off" position.*

Once stored energy has been released, the equipment has been tested to verify its isolation, and the operating controls have been returned to the "off" position, then your equipment is locked out or tagged out. Let the repairs begin!

Regulatory Code:

> 1910.147(e) Release from lockout or tagout. Before lockout or tagout devices are removed and energy is restored to the machine or equipment, procedures shall be followed and actions taken by the authorized employee(s) to ensure the following: (1) The machine or equipment. The workarea shall be inspected to ensure that nonessential items have been removed and to ensure that machine or equipment components are operationally intact. (2) Employees. (i) The workarea shall be checked to ensure that all employees have been safely positioned or removed. (ii) Before lockout or tagout devices are removed and before machines or equipment are energized, affected employees shall be notified that the lockout or tagout devices have been removed. (3) Lockout or tagout devices removal. Each lockout or tagout device shall be removed from each energy isolating device by the employee who applied the device. Exception: When the authorized employee who applied the lockout or tagout device is not available to remove it, that device may be removed under the direction of the employer, provided that specific procedures and training for such removal have been developed, documented and incorporated into the employer's Energy Control Program. The employer shall demonstrate that the specific procedure provides equivalent safety to the removal of the device by the authorized employee who applied it. The specific procedures shall include at least the following elements: (i) Verification by the employer that the authorized employee who applied the device is not at the facility; (ii) Making all reasonable efforts to contact the authorized employee to inform him that his lockout or tagout device has been removed; and (iii) Ensuring that the authorized employee has this knowledge before he resumes work at the facility.

English: Your energy control procedures are not complete until you have determined how authorized employees will release machines and equipment from lockout or tagout and reenergize them after they have finished their maintenance or repairs. In other words, how are they going to set up and return to normal operations?

Returning to Normal

Unlike procedures for shutdown, energy isolation, and placement of locks and tags, which must be equipment-specific, these procedures apply to any machinery, equipment, or size facility. You can set one procedure to be followed plant-wide. The Standard identifies three key elements to safe return of equipment to normal operations:

1. Locks and tags must be removed by the person(s) who placed them.
2. The workarea must be inspected.
3. Affected employees must be notified.

Locks and Tags Must Be Removed by the Person Who Placed Them
This little statement must reside at the core of your Energy Control Program. The purpose of all of these procedures is, after all, to provide the *maximum level of control* over energy hazards to persons conducting maintenance and servicing. The reason for using locks or tags in the first place is to prevent someone from walking by and flipping a switch, plugging in a cord, or otherwise operating machinery or equipment while it is being worked on. If persons other than those who place locks and tags are allowed to remove locks and tags, the level of protection and security provided to the person doing the servicing is greatly diminished.

This brings up the importance of training everyone in your company in the purpose of locks and tags and in the proper procedures for their removal. If no one else at your company knows anything at all about locks and tags except your maintenance crew, then the potential exists, especially where tags are used, for another employee to come across a tag and think (or *not think*, as the case may be), "Hmm, never seen this before. Must be some sort of manufacturer's warning." And simply ignore it, bypass it, or remove it. Especially at large operations where everybody doesn't know what everybody else is doing or where complex, multi-workstation equipment is used, lack of knowledge of the purpose of locks and tags and of safe removal procedures *by everyone at the facility* can be disastrous.

Group Lockout or Tagout
When group lockout or tagout is performed, as might be the case when several maintenance personnel are repairing an air conditioning system, procedures for

removal of locks and tags must include the provision that the equipment will not be reenergized until everyone is out of the danger zone. According to paragraph (f) of the Standard, this can be accomplished in one of two basic ways:

1. By designating an operation leader to attach a single group lock or tag to the appropriate energy isolating device(s) and requiring each person involved in the maintenance to check in with this leader when they have completed their work. The operation leader is responsible for making sure that everyone is safe before removing the group lock or tag.
2. By requiring each person involved in the group maintenance to attach his or her own personal lock or tag to the energy isolating device in such a manner that it cannot be operated until all locks or tags have been removed.

In all cases of group lockout or tagout, each individual involved must either personally check in with the designated leader or personally remove their own lock or tag before equipment is reenergized. In dealing with the kind of large electrical, mechanical, and chemical processes that require group maintenance, it is important never to rely on hearsay in deciding whether or not it is safe to reenergize. Bob may very well have seen Steve follow him out of a danger zone two minutes ago. What Bob may not have seen was Steve reentering the danger zone to retrieve a dropped wrench.

The locks and tags used at your facility are safety devices in much the same way that respirators, goggles, and hardhats are safety devices. In order to insure that they command the respect that they should around your plant, your authorized employees must use them correctly.

For example, when an authorized employee returns an energy isolating device to the "on" or "operating" position, he needs to remove the lock or tag from the device. He should never leave a lock or tag hanging from a circuit breaker that is "on" or a valve that is "open." Why? Because this sends a mixed message to the affected employees, his fellow workers, whom he wants to abide by the lock or tag. If a tag that says "Do Not Operate" is left hanging from a machine in obvious operation, it is sending the direct message to everyone in the area: *Ignore this tag, it means nothing.* In order for affected employees, which is everyone at your company, to respect that tags and locks really do mean "Do Not Operate," they must be used *and removed* responsibly.

Exception to the Removal Rule
In paragraph (e)(3), the Standard does allow that there may be times when it is simply not possible to have the employee who placed a lock or tag remove it, such as if they become seriously ill or are dismissed from work while their equipment is under lock or tag. It is important to keep in mind that the inclusion of this allowance should not be read as license to loosen up on your removal procedures.

Additionally, the removal of a lock or tag and subsequent reenergization of a machine by someone other than the person who isolated it requires, above all, that this person has the necessary training and knowledge of the machine to ensure an equivalent level of safety as would have been provided by the person who originally placed the lock or tag.

The only time that removal of a lock or tag by someone other than the person who placed it should be allowed at your facility is when these four conditions have been met:

1. You have verified that the employee is not at the facility.
2. You have made all reasonable efforts to contact the employee.
3. You have insured that the employee who placed the lock or tag will be informed of its removal before he or she resumes work.
4. You have insured that the employee removing the lock or tag has sufficient knowledge to protect his or her own safety and that of affected employees.

The Workarea Must Be Inspected
Always, before authorized personnel reenergize or turn on equipment after conducting maintenance or repairs, they need to inspect both the *equipment itself* and the *area around it*. This is a simple standard operating procedure that can help avoid damage to equipment and tools, as well as the creation of additional safety hazards.

You might want to develop the procedure as a checklist that can be quickly reviewed after all maintenance or servicing activities:

1. Have all tools been removed?
2. Is anything missing from your tool chest, cart, etc.?
3. Has equipment been put back together correctly?
4. Have any parts been left out?
5. Are all guards and safety devices in place?
6. Have all extraneous items (coffee cups, Coke cans, paper) been removed?
7. If any spills or leaks occurred, have they been cleaned up?

Affected Employees Must Be Notified
Always, before authorized personnel reenergize or turn on equipment after conducting maintenance or repairs, they need to notify everyone who operates the equipment or works in the general vicinity of the equipment that lockout or tagout devices have been removed and that the equipment is going to be turned back on.

Before actually reenergizing the equipment, they need to check the area around the equipment carefully to make sure that everyone is safely positioned. Depending on the size of the equipment, you may want to include in your procedures the use of an "All Clear" verbal warning or other alarm to indicate imminent start-up. This is an important procedure to remember in the context of reenergization after an emergency shut down. Although locks and tags would not be used in such a circumstance, the personnel conducting the start-up would need to take the same precautions to insure the safety of other employees in the workarea.

Regulatory Code:

1910.147 (6) Periodic inspection. (i) The employer shall conduct a periodic inspection of the energy control procedures at least annually to ensure that the procedure and the requirements of this Standard are being followed. (A) The periodic inspection shall be performed by an authorized employee other than the one(s) utilizing the energy control procedure. (B) The periodic inspection shall be designed to correct any deviations or inadequacies observed. (C) Where lockout is used for energy control, the periodic inspection shall include a review, between the inspector and each authorized and affected employee, of that employee's responsibilities under the energy control procedures and the elements set forth in paragraph (c)(7)(ii). (ii) The employer shall certify that the periodic inspections have been performed. The certification shall identify the machine or equipment on which the energy control procedure was being utilized, the date of the inspection, the employees included in the inspection, and the person performing the inspection. (7) Training and communication (i) The employer shall provide training to ensure that the purpose and function of the Energy Control Program are understood by employees and that the knowledge and skills required for the safe application, usage and removal of energy controls are required by employees. The training shall include the following: (A) Each authorized employee shall receive training in the recognition of applicable hazardous energy sources, the type and magnitude of the energy available in the workplace, and the methods and means necessary for energy isolation and control. (B) Each affected employee shall be instructed in the purpose and use of the energy control procedure. (C) All other employees whose work operations are or may be in an area where energy control procedures may be utilized, shall be instructed about the procedure and about the prohibition relating to attempts to restart or reenergize machines or equipment which are locked or tagged out. (ii) When tagout systems are used, employees shall also be trained in the following limitations of tags: (A) Tags are essentially warning devices affixed to energy isolating devices, and do not provide the physical restraint on those devices that is provided by a lock. (B) When a tag is attached to an energy isolating means, it is not to be removed without authorization of the authorized person responsible for it, and it is never bypassed, ignored, or otherwise defeated. (C) Tags must be legible and understandable by all authorized employees, affected employees, and all other employees whose work operations are or may be in the area, in order to be effective. (D) Tags and their means of attachment must be made of materials which will withstand the environmental conditions encountered in the workplace. (E) Tags may evoke a false sense of security, and their meaning needs to be understood as part of the overall Energy Control Program. (F) Tags must be securely attached to energy isolating devices so that they cannot be inadvertently or accidentally detached during use. (iii) Employee retraining. (A) Retraining shall be provided for all authorized and affected employees whenever there is a change in their job assignments, a change in machines, equipment or processes that present a new hazard, or when there is a change in the energy control procedures. (B) Additional retraining shall also be conducted whenever a periodic inspection under paragraph (c)(6) of this section reveals, or whenever the employer has reason to believe that there are deviations from or inadequacies in the employee's knowledge or use of the energy control procedures. (C) The retraining shall reestablish employee proficiency and introduce new or revised control methods and procedures, as necessary. (iv) The employer shall certify that employee training has been accomplished and is being kept up to date. The certification shall contain each employee's name and dates of training.

English: Who needs training under this Standard? And what must it cover?

Developing a Training Program

Paragraph (c)(7)(i) indicates three categories of employees who need to be trained in the energy control procedures at their company. For simplicity sake, that number can be reduced to two: authorized employees and affected employees.

Authorized Employees

As mentioned earlier, authorized employees are those who will be conducting maintenance or servicing activities under lockout or tagout. At your facility, authorized employees may be limited to your maintenance staff, they may include selected equipment operators or supervisors, or they may include everyone on your production floor. You may decide not to authorize any of your employees to conduct maintenance or repairs requiring lockout or tagout, but to call in outside contractors to do that work. Who you decide to authorize is entirely up to you. OSHA just requires that you document your decision and develop procedures to enforce it.

It is for these authorized employees that you develop the five-step energy control procedure outlined above. What does their training need to cover? Exactly those five steps, plus the procedures for releasing machines and equipment from lockout or tagout. As you review the elements of your energy control procedure, simply ask yourself what authorized employees need to know.

First, they need to know who they are. They need to know that they are authorized to use lockout and/or tagout.

Second, they need to know when energy hazards are present and the type and magnitude of that energy. For each machine applicable to them, they need to know during which specific maintenance or servicing activities lockout or tagout is required to control that energy.

Third, they need to know how to isolate energy from each machine or equipment they work on. They need to know where the energy isolating devices for each machine are physically located, as well as how to operate them.

Fourth, they need to know which equipment can and cannot accept locks, how to use locks, when tags are acceptable, how and where tags must be attached to energy isolating devices, and the limitations of tags.

Fifth, they need to know which machines or equipment store hazardous energy after shutdown and isolation and how to release or block that energy. They also need to know how to verify that a machine has been effectively isolated before they begin maintenance or repairs.

Sixth, they need to know how to safely release the machines or equipment they work on from lockout or tagout and return them to normal operations.

In addition to classroom discussion, including review of the written program, energy isolating device charts, and locks and tags, training for authorized employees must involve physically walking around the plant, identifying where and how to attach locks and tags, and going through the motions of a complete energy control procedure, from preparation for shutdown to release from lockout or tagout. In order to effectively prepare your authorized personnel to safely use locks and tags, training must emphasize hands-on practice. An added benefit of this approach is that it allows you the opportunity to test your procedures in a controlled atmosphere, see if they work, and modify, trim, or expand them as required. As discussed in Chapter 5, this proactive approach to training allows employees to be involved in the development of the procedures they will be expected to follow.

Affected Employees

Although the Standard makes a vague distinction between affected employees and a third category, all other employees who may be in an area where energy control procedures are utilized, it is more practical to lump these two groups together. In other words, all employees who are not authorized to perform lockout or tagout should be considered affected by lockout or tagout. Why? Because it is virtually impossible to set hard and fast rules regarding who at your company may or may not be affected by energy control procedures.

While it is clear that operators of energized equipment are affected by energy control procedures, since lockout or tagout may be performed on the equipment they operate, it is less clear when materials handlers, janitors, truck drivers, lab technicians, managers, or office staff may be working in an area where energy control procedures are being utilized. Bottom line, the safest policy is to treat *every employee* as an affected employee.

What do affected employees need to know about energy control procedures? They need to know enough to insure that they do not interfere with or create any obstacles to an authorized employee carrying out the procedures.

Specifically, they need to know the following information, as it relates to your facility:

- They need to know who is authorized to use locks and tags and why they use them.
- They need to be able to recognize the locks and tags used by authorized employees.
- They need to know never to remove locks or tags or to attempt to energize machines or equipment that are locked or tagged.
- They need to know that tags are only warning devices that provide no physical restraint against the start-up of equipment.
- They need to be instructed always to report locked or tagged equipment to their supervisor rather than to attempt to remove or bypass the device themselves.

The purpose of this training is to build awareness of the Energy Control Program at your company and to foster respect, in every employee, for the lockout and/or tagout procedures followed by authorized personnel.

When do employees have to be trained? And how should it be documented?

Documenting Training
Obviously, everyone needs to be trained at the time the energy control procedures are developed. After procedures are in place, new employees must be trained at the time they are hired. At that initial training, a record needs to be established indicating the name of the employee trained, the date, and the topics covered. Although the Standard does not specifically require employee testing, using a written test is an excellent method for "certifying" that employees have been trained. Site-specific multiple choice or true/false tests are simple to develop and serve as one indicator of an employee's understanding of the concepts and procedures covered in training (see Figure 6-4).

The Standard requires retraining whenever there is a change in job assignments, machines, equipment, or processes, or whenever the procedures themselves are changed. Retraining must also be provided whenever an inspection indicates that employees are deviating from the established energy control procedures or simply don't know enough about the procedures to follow them. In making this connection between inspections and the need for additional training, the Lockout/Tagout Rule provides employers practical and highly specific guidance that is absent in most training regulations.

In paragraph (c)(6) the Standard requires that employers conduct and document periodic inspections of the energy control procedures, and thus the effectiveness of training, by walking around the production floor and talking to authorized and affected employees about Lockout/Tagout. Inspections should include asking employees questions such as: *When do you lockout this equipment? Where do you attach the lock? How do you verify that the equipment is isolated? Is there any stored energy that needs to be released? How do you release it? If a lock is on this circuit breaker, what does that mean? If a "Do Not Operate" tag is on this switch,*

LOCKOUT/TAGOUT EMPLOYEE EVALUATION

Name: _____ Date: _____

1. The purpose of lockout procedures is to insure that equipment won't be energized while it is being serviced.   TRUE OR FALSE

2. When operating a press, injuries can be prevented by keeping guards in place.   TRUE OR FALSE

3. If I have locked or tagged out a piece of equipment, before I operate it again, I should:
    a. Check the area around the equipment to make sure no one is at risk.
    b. Remove tools from equipment.
    c. Replace guards on the equipment.
    d. All of the above.

4. When I place a "DO NOT OPERATE" tag on equipment, the equipment can't be operated.   TRUE OR FALSE

5. When I lock the circuit breaker for a machine in the "off" position, the machine can't be operated.   TRUE OR FALSE

6. In the event that I should have to evacuate the building, I should turn off my machine before I leave.   TRUE OR FALSE

7. If taking a short cut means increasing production, then my supervisor wants me to take a short cut, even if I compromise my own safety.   TRUE OR FALSE

8. As long as I have at least three years experience, it is safe to operate energized equipment when I am alone in the shop.   TRUE OR FALSE

9. At the first sign of equipment malfunction, I should:
    a. Ignore it and keep working.
    b. Work faster to try to get the job done.
    c. Stop and notify my supervisor.

10. When repair personnel are doing electrical work or replacing parts, they will lock out the equipment or use tags when equipment can't be locked.   TRUE OR FALSE

Figure 6-4. Excerpt from Model Lockout/Tagout Employee Evaluation.

what does that mean? If you did not place a "Do Not Operate" tag, is it okay to remove it?

Inspections should also include physical examination of energy isolating devices to make sure that they can be and are being locked or tagged in accordance with established procedure and to make sure that locks and tags are being removed completely from energy isolating devices when equipment is returned to normal operations. Documentation of inspections should include the name of the inspector, the name(s) of the employee(s) interviewed, the equipment or procedures discussed, and the date. Although the Standard does not require this, it makes sense to include on your inspection report the results of the inspection or interviews. The report then can serve double duty as a training format or outline upon which you can base and document your retraining.

The Standard requires retraining only when equipment or procedures change or when inspections indicate deficiencies. However, as with the Hazard Communication Standard, it makes good sense to conduct retraining at least annually, regardless of facility changes or the results of inspections. Annual training must be a minimum training policy if employee safety and health is to be served.

Assessing the Program

Of all of the questions asked by employers or managers facing compliance with an OSHA Standard, these are the most common: *How do I know if I'm doing it right? How do I know if my training is successful? How do I know if I'm meeting the requirements of the Standard?*

Remember, OSHA Standards are performance-oriented; their sole purpose is to *influence behavior* so that the workplace improves, becoming a safer and healthier place to work. While written and verbal tests serve as one good indicator of basic knowledge, the only way to find out whether or not your employees can apply that knowledge is to *watch them in action.* Inspections will let you know whether or not your training is effectively influencing behavior, or the way your employees do things in the shop. If Lockout/Tagout inspections indicate that employees are not modifying standard operating procedures to incorporate energy control procedures, then your training is not working.

If this is the case, what do you do? Well, depending where the problem lies, you can try several things.

If, by talking to your employees, you discover that many lack understanding of Lockout/Tagout procedures or think that the procedures are silly or a waste of time, then additional classroom discussion and hands-on practice implementing the procedures might be in order. *More training and information* may be all that is needed to influence the behavior of your employees.

Or, perhaps some of your authorized employees see problems in the Lockout/Tagout procedures that they haven't told you about, choosing instead to adopt an arrogant attitude about procedures they "know won't work." For instance, if you have told them to lock out a machine during a certain operation and they know full well that the machine *can't* be locked out during that operation because they must be able to jog it to conduct the repairs, then, obviously, your procedure is worthless. When, during an inspection, you discover that your procedure isn't being followed, instead of demanding that they follow it, first ask them *why they aren't following it.* Break through the communication barrier that keeps employers and employees at opposite poles by picking their brains. Encourage them to dissect the Lockout/Tagout procedures. Challenge them to make the procedures better. *Accurate, practical procedures* may be all that your employees need in order to change the way they do things.

As mentioned in Chapter 4 in the context of personal protective equipment, perhaps your employees lack proper incentive to incorporate Lockout/Tagout procedures into their standard operating procedures for maintenance and servicing. If energy control procedures are not company policy and they are not enforceable by disciplinary action, if, when your inspections reveal blatant disregard of the procedures you just shrug or look the other way, or if, when production schedules get tight, you allow or even encourage safety procedures to be bypassed or ignored, then why in the world should your employees choose to follow them? The understanding that following Lockout/Tagout procedures is company policy and that failure to do so will result in *disciplinary action* may be all the incentive your employees need to alter their behavior.

Training Contract Employees

Regulatory Code:

> 1910.147 (f)(2) Outside personnel (Contractors, etc.). Whenever outside servicing personnel are to be engaged in activities covered by the scope and application of this Standard, the on-site employer and the outside employer shall inform each other of their respective lockout or tagout procedures. (ii) The on-site employer shall ensure that his or her personnel understand and comply with restrictions and prohibitions of the outside employer's energy control procedures.

English: Who is responsible for outside personnel?

Let's say you have equipment that is connected to some source of energy (electrical, mechanical, hydraulic, pneumatic, chemical, or thermal). All the proper guards are in place. Your operators have been trained never to bypass a safety device or stick a body part into a danger zone. Whenever maintenance that would call for the removal of a guard or safety device is required, you call an outside contractor. Does this mean you do or don't have to have an energy control procedure?

The language of the Standard makes the clear assumption that both the on-site employer and the contractor will have their own procedures. As the on-site employer, when contractor employees use locks and tags at your facility, it is your responsibility to make sure that your employees understand the procedures used by the contractor to the extent that they *don't interfere with them.*

If all employers had their own energy control procedures, this system would work just fine. However, what do you do if a contractor employee begins work at your facility without using *any* lockout or tagout procedures? You could just let them take their chances—after all, the Standard doesn't directly state that if a

136   Safety and Environmental Training

contractor doesn't have an energy control procedure, the on-site employer must provide one. But would allowing such risk to be taken at your facility be wise? Probably not—especially if your employees might be affected by it. As a rule of thumb, you should never let anyone conduct servicing or repairs on energized equipment unless lockout or tagout procedures are followed. If you want to save yourself some work, you might inform your contractors that having an energy control procedure is a condition of keeping their contract.

Exception to the Requirement for a Written Energy Control Procedure

Regulatory Code:

> 1910.147 (c)(4)(i) Note: The employer need not document the required procedure for a particular machine or equipment, when all of the following elements exist: (1) The machine or equipment has no potential for stored or residual energy or reaccumulation of stored energy after the shutdown which could endanger employees; (2) The machine or equipment has a single energy source which can be readily identified and isolated; (3) the isolation and locking out of that energy source will completely deenergize and deactivate the machine or equipment; (4) the machine or equipment is isolated from that energy source and locked out during servicing or maintenance; (5) a single lockout device will achieve a locked out condition; (6) the lockout device is under the exclusive control of the authorized employee performing the servicing or maintenance; (7) the servicing or maintenance does not create hazards for other employees; and (8) the employer, in utilizing this exception, has had no accidents involving the unexpected activation or reenergization of the machine or equipment during servicing or maintenance.

English: Hold your horses. Before you start celebrating your escape from all the work outlined in this chapter, there are a couple of things you need to understand.

First, you need to understand that this exemption applies only to machines or equipment that can be locked out and completely deactivated by the application of a *single lock*.

If your equipment cannot be locked out or if you choose to use a tag rather than a lock or you would like to have the *option* of using a tag rather than a lock, this exemption does not apply to you.

Second, you need to know that if all eight conditions listed in paragraph (c)(4)(i) apply to a machine or equipment at your company, then you *still must develop a lockout procedure* for that machine or equipment. This paragraph exempts you only from having to write the procedure down on paper.

So, if you have a machine that is locked out with a single lock that is under the control of the person conducting the servicing, that is completely deactivated by this lockout and stores no energy or can regenerate no energy, that creates no hazards for other employees while it is being serviced, and that has caused no accidents or injuries through unexpected energization during maintenance or servicing, then you do not have to develop a written energy control procedure for that particular machine.

You do, however, still have to train your authorized and affected employees in an energy control procedure that includes preparation for shutdown, shut down, isolation, lockout device application, verification of isolation, and release from lockout. If you can verify that a given machine meets the conditions for this exemption and insure that your employees are sufficiently trained in this procedure *without putting anything on paper,* then more power to you.

One word of warning: It is difficult to prove that your employees have been trained in a procedure that is only word of mouth and it is even harder to enforce a policy that has never been written down. It appears that this exemption more than makes up, in headaches and risk, for whatever work it may initially "save" you.

USING STANDARD OPERATING PROCEDURES

While the purpose of the Lockout/Tagout Rule is to provide procedures to avoid the unexpected energization of equipment during servicing and maintenance, with a little imagination you will see how the energy control procedures developed to comply with this Standard can be used to accomplish other things at your workplace. Start by thinking of the energy control procedures for each piece of equipment at your company as utterly inseparable from standard operating procedures. What are the advantages of making energy control procedures part of the bigger picture of your standard operating procedures for both equipment maintenance and equipment operation?

Looking back at the beginning of this chapter, it is mentioned that hazard identification must be one stage in the development of an energy control procedure. In identifying energy hazards, both routine and non-routine maintenance and servicing activities must be identified and examined.

Operator Maintenance

Routine maintenance and servicing, according to the Lockout/Tagout Rule, are the lubrication activities and small tool adjustments that an equipment operator must complete in order to keep his or her equipment running. In developing a standard operating procedure for these activities, you might call them Operator Maintenance. This procedure should include everything the operator needs to do, inspect, or monitor on his or her equipment (and the intervals at which the activities should be done), to maximize the possibility of intercepting minor problems before they become serious and to minimize sudden breakdowns. For each activity, the standard operating procedure should include any personal protective equipment (gloves, glasses, goggles, respirator) that the operator needs to wear.

For Operator Maintenance that *is exempt* from requiring lockout or tagout, standard operating procedures must specify that machine guards and safety devices cannot be removed. You need to indicate where the operator can and cannot put his or her hand or any other part of the body. You also need to indicate any particular procedures that are *not* allowed, such as cleaning a moving roller or gear and using compressed air to blow grease or dirt out of an equipment part. For Operator Maintenance that *is not exempt* from lockout or tagout, the specific locking or tagging procedure needs to be described, as outlined earlier in the chapter.

By combining energy control procedures with standard operating procedures for Operator Maintenance, what have you gained?

1. You have formalized a procedure for maintaining your equipment or preventing it from breaking down. In other words, *maintaining* equipment has been established as part and parcel of *operating* equipment.
2. You have formalized the expectation that these maintenance activities will be done *in accordance with safety procedures*.
3. You have indicated when lockout or tagout is required and have established that lockout or tagout is inseparable from certain maintenance activities integral to the operation of equipment.
4. You have a single, coherent procedure from which to train.

Another category of routine maintenance and servicing involves emergency and fire prevention equipment. As explained in Chapter 5, routine inspection, monitoring, and repair of emergency and fire prevention equipment must be part of your OSHA Employee Emergency Plan and Fire Prevention Plan as well as your RCRA Contingency Plan. Standard operating procedures for the maintenance of this equipment would include a maintenance schedule, a description of the inspection, monitoring or servicing to be conducted, and an explanation of the safety procedures to be followed. If lockout or tagout is not required, the procedures should include special safety precautions that the person conducting the maintenance needs to follow, including the equipment that needs to be turned off, the equipment that cannot be turned off, areas where hands or other parts of the body should not be placed, and the protective equipment that must be worn. For servicing of emergency or fire prevention equipment that requires lockout or tagout in order to insure the safety of employees, the specific locking or tagging procedure needs to be described.

What are the benefits of establishing one standard operating procedure for the maintenance of emergency and fire prevention equipment?

1. You have formalized a procedure and schedule for maintaining your emergency and fire prevention equipment so that it will always be ready and working when it is needed.

More Than Maintenance   139

2. You have formalized the expectation that these maintenance activities will be done in accordance with safety procedures.
3. You have indicated when lockout or tagout is required and have established that lockout or tagout is inseparable from certain maintenance activities integral to the continuous operation of emergency and fire prevention equipment.
4. You have a single, coherent procedure from which to train.

Non-Routine Maintenance and Repairs

Non-routine maintenance activities are, generally speaking, those that are conducted out of the context of daily production operations. These activities include major cleaning, lubrication, replacement of parts, rewiring and unjamming of machines or equipment, and always require lockout or tagout. Although it varies from company to company, nonroutine maintenance generally is performed by either designated maintenance personnel or by outside contractors.

As with Operator Maintenance, the purpose of combining your established maintenance standard operating procedures with your Lockout/Tagout procedures and calling the whole thing your Standard Operating Procedures for Preventive Maintenance and Equipment Repairs (or something similar) is to make the point that safety procedures cannot be separated from the maintenance activities themselves. Perhaps it's more a matter of semantics than anything else, but if you present a new program to your maintenance staff as "Revised Standard Operating Procedures for Preventive Maintenance" rather than as "New Safety Procedures," it might be taken a little more seriously.

By using the standard operating procedure, you can cover other safety precautions that must be taken during maintenance activities. You are not limited only to those that directly concern energy control. For instance, it is something of a waste to have a maintenance standard operating procedure that includes lockout procedures on chemical input valves, yet doesn't mention the protective equipment that the technician should use for protection from the chemical. In order to be optimally effective, standard operating procedures for maintenance activities must include *all* procedures relevant to the safety of the employee performing the servicing, whether they include chemical hazards, protective equipment, or emergency procedures.

The Lockout/Tagout Rule provides a limited but focused place at which to begin combining standard operating procedures with safety procedures. But it is a very important place, because an effective Operator, Emergency Equipment, and Preventive Maintenance program is critical to a safe workplace. As you have seen, even simple maintenance activities overlap with several safety issues, providing an opportunity to establish one procedure that includes them all.

7

Before It Breaks, Fix It!

For here we are not afraid to follow truth
Wherever it may lead, nor to tolerate error
So long as reason is free to combat it.

Thomas Jefferson

It's unavoidable. Things are going to have to change. Out with the old; in with the new. We're talking major spring cleaning here. And no one or nothing, not the President of the company, the employee who's been with you the longest, or the oldest piece of equipment on the production floor, is immune. In order to meet the expanding demands of safety and environmental regulations, the standard operating procedures at your company are going to have to be replaced with a newer model.

Instead of waiting for the old standard operating procedures to fail you, *reinvent them now.* How? *Incorporate your training content into your standard operating procedures.* And then (and this is the good part) present your training content not as safety procedures or environmental compliance, but as *revised standard operating procedures.*

Ah, we may be onto something here. As we saw in Chapter 6, within the limited context of Lockout/Tagout, by revising your standard operating procedures to include safety and environmental concerns, you accomplish several things:

1. You cover all the regulatory requirements that impact a certain activity *along with* the operating or production requirements for that activity. This allows you to present one coherent procedure to your employees and label it "standard."
2. You establish that the company "standard" is to operate in a safe and healthy manner. You make it loud and clear that safety and environmental concerns cannot be separated from production operations.
3. You get your employees' attention. If there is a tendency at your company for folks to follow safety procedures only when it's convenient or when production speed isn't compromised, presenting safety not as an optional accessory external

to production, but as an essential component of standard production operations, might prompt them to re-think their attitude.

Far too often, employees are given several procedures to follow: one for production (which is known to be the most important); one for safety (which is often in direct contradiction to the production standard operating procedure); and perhaps yet another one for environmental compliance purposes. When this happens, confusion and resentment result. And more often than not, safety *and* compliance suffer.

Generally, workers will identify the standard operating procedure for their job as the way they were taught to do it or the way they have observed their coworkers doing it. This procedure, regardless of its incorrectness or unsafeness, becomes the "standard" or ethic for the job. From the worker's perspective, to deviate from the "standard" would be, at the very least, to risk ridicule for going against the grain, and, at most, to risk disciplinary action, perhaps even being fired, for changing an established, tried and true procedure. Often, simple habit rather than some proven methodology plays the largest role in shaping what can evolve into monolithic and seemingly unchangeable standard operating procedures.

Case in Point: All the workers in the painting department at a manufacturing company get in the habit, during a particularly busy production period, of using lacquer thinner instead of soap to clean paint off their arms and hands. The Production Supervisor doesn't say anything to them about it, figuring that using the thinner is a lot faster than cleaning with soap, allowing the men to work a little longer before lunch and up until quitting time. The thinner washing continues. Soon soap is abandoned altogether. A year later the men have forgotten they ever used soap. New hires are taught thinner-washing as the standard operating procedure. When the Safety Manager talks to the workers about health hazards and protective equipment at a Right-to-Know class, his pleas for them to wear gloves and to start using soap fall on deaf ears. Washing with thinner has become so much a part of the way they do things that, although what the Safety Manager is saying *sounds* good, they can't see how it *applies to them.* They can't imagine doing their job without washing with thinner.

DEFINING THE SAFE OPERATING PROCEDURE

Traditionally, a standard operating procedure consists of a narrative, step-by-step description of how a job is conducted, equipment is run, a product is made, or a service is rendered. Sometimes standard operating procedures are written and sometimes they are communicated verbally. Standard operating procedures are what a new employee learns during on-the-job training. They are the tried and true methods for doing everything at your company that enables you to manufacture a product, provide a service, and meet your payroll.

Since standard operating procedures cover things like how fast a machine can run, how much pressure can be built up in line, which size wood fits through a saw, how much paper a cutter can cut, how often oil needs changing in a forklift, how to bleed pressure off a tanker truck, and how to drain a fuel transfer hose and where to store it, the tendency is to think of standard operating procedures as entirely production-oriented procedures—the singular domain of the Production Manager and Supervisors, having nothing whatsoever to do with safety or environmental compliance.

Safety procedures are more likely viewed as being entirely separate from the procedures that must be followed to make a product or provide a service. Safety procedures are what people follow only if they don't interfere with standard operating procedures, if they don't cramp the accepted work style or "waste" any time. As soon as standard operating procedures are compromised or the speed at which employees can accomplish a task is diminished, then safety procedures are tossed out the window.

Traditionally, standard operating procedures are what employees follow *or else they're fired.* The procedures they hear about in safety and environmental training classes are what employees follow *as long as standard operating procedures are not affected.*

How does this impact the workplace?

When an employer fails to present safety procedures as part of standard operating procedures or as integral to the production of a quality product, then the employer knowingly or not, encourages the attitude that, come hell or high water, the only thing that matters is meeting the quota and getting the job out—no matter who or what gets lost in the crush to produce.

In other words, the message communicated is: *safety is dispensable, and therefore* (to a greater or lesser extent) *so are you.*

As discussed in depth in Chapter 8, the primary reason why the dichotomy between standard operating procedures and safety procedures exists is because safety (or compliance with safety and environmental regulations) is not viewed by the person writing the standard operating procedures, the person writing the safety procedures, or the employer as a *production concern.*

The myth persists that compliance can be achieved without altering "the way things are done." This is impossible. In order to meet the intent of safety and environmental regulations, the protection of employees and the environment must be viewed as an essential aspect of, *not a detriment to,* efficient production. However, once this connection is made, it becomes obvious that writing standard operating procedures that don't include safety procedures makes about as much sense as buying a sports car without brakes.

Whereas the traditional standard operating procedure is like an heirloom passed down from generation to generation, unchanged, monolithic, and becoming more precious with age, this new brand of standard operating procedure—what you might think of as a *safe operating procedure*—is anything but carved in stone. This safe operating procedure, which is born in a training class, reviewed in production meetings, and practiced on the production floor, is routinely recreated, revised, and rediscovered. Because it is a procedure that incorporates many aspects of being safe while working, it makes sense that as regulations change, employees change, and equipment changes, it too will need to grow and change.

You can spend a lot of money on your safety and environmental training program. You can call in experts to help you design it. You can use state of the art resource materials, flashy videos, colorful comic books, and slides. You can buy the best equipment. You can even spend more hours in the breakroom or the conference room talking about safety and health than you spend every year on vacation.

And after you've done all of that, after you've worked and sweated and tried real hard, unless day-in day-out procedures change, you will have bought yourself the equivalent of a finely tuned, exquisitely painted Ferrari that doesn't have any tires—the day after every tire in the world disappears.

Your safety and environmental program looks pretty. Sounds even better. But, when you get right down to it, it doesn't amount to a hill of beans.

What would you think of a football coach who spent hours upon hours with his players, going over plays, memorizing strategies and statistics, and analyzing films of victories and defeats. Good coach, huh? Smart guy.

How would your opinion change if you found out that this coach never gave his team the ball? He never let them out on the field. Never let them play a game. Sure, they were brilliant in the classroom, on paper, in theory, but the team never was required to put knowledge into practice, to bring dry theory to sweaty life. Their knowledge was important only within the confines of the classroom, *for its own sake* and *with no practical purpose,* since this coach never let his team stand in the glare of the stadium lights and move the football down the field.

When you train to comply with the law and not to change procedures at your company—no matter how much money, time, and effort you spend doing it—in the end you're just like that coach, or the owner of a Ferrari with no tires.

SET GOALS, GET RESULTS

"But how do I know if I am really in compliance?"
"How do I know if I'm giving OSHA what they really want?"
"How can I be sure that my training program is adequate?"

If you are responsible for compliance with training regulations, these questions no doubt have crossed your mind. Perhaps you have fantasized about a surefire formula for success, some document or person that could give you the key to the elusive mind of OSHA or EPA...if only you could find it.

Performance-oriented Standards, such as Hazard Communication and Lockout/Tagout, are Standards for which the goals are to change behavior in employees. There are no miracle cures for compliance. Change in this context, as in all others, is a gradual, often arduous process. Too often however, the tendency is to take a checklist approach to compliance with these laws. A checklist approach is a minimalist approach that fosters the attitude: "How can I comply with the letter of the law and spend as little time and money as possible?" As attractive as this approach might seem, it directly contradicts the intent of performance-oriented standards, making compliance through this method as verifiable as a myth.

The Myth of the Checklist Approach

Consider compliance with Right-to-Know. What could be easier than showing two 10-minute videos on Material Safety Data Sheets and labeling, identifying where the Material Safety Data Sheets for the facility are kept, and having employees sign a training record that says they have been told they have a "right-to-know" about the chemical hazards in the workplace?

It's cheap. Videos cost from $250 to $600 and can be shown again next year. It's short. The whole process takes only 30 minutes. It's sweet. Signed records are in the file, ready to impress the next OSHA inspector or insurance auditor who walks through the door.

But will they?

That depends on what the inspectors are looking for. If they are looking for signed attendance records, then you've got it made. But what happens if they go out on your production floor and start asking your employees questions about the hazards of their jobs? *What will your employees say?* Or, if they walk around the workplace to observe your employees' chemical handling procedures, *what will they find?* It is unlikely that a written program on a shelf in your office or that 10-minute video shown a year ago will help your line operator, welder, or pressman answer questions about hydraulic oil, compressed gases, or blanket wash. And if chemical safety and protective equipment have not been integrated into your employees' safe operating procedures, then is it likely the inspectors will find any evidence of compliance as they watch your people work?

Regarding Lockout/Tagout, the same holds true. You can show a video, fill a drawer with locks and tags, and have your employees sign forms saying they've been trained. Seems simple enough. But if the purpose of Lockout/Tagout is to raise awareness and foster understanding of how to avoid accidents while working with energized equipment, what good will any of these *activities* by themselves actually accomplish?

Locks and tags in a drawer, while they may (or may not) get you through an inspection, will not raise awareness. If an inspector asks a machinist to describe those maintenance activities that require lockout or tagout procedures, is it likely that the machinist will have learned these procedures from a video? When you put locks and tags in a drawer without holding a class to explain their purpose, or when you show videos without asking employees to apply the information presented to their particular jobs, you take a checklist approach to compliance. You are doing what the regulation says to do, but you're not making those activities have any impact at your facility. In your training, you may be accurately, even masterfully, reiterating the regulations, but if training doesn't incorporate safe operating procedures, you won't change behavior. You won't see results.

When you take a checklist approach to compliance, safety and environmental regulations are obscured and trivialized. Instead of being used as a management tool that strengthens and unites your workforce by teaching its members how to care for themselves, compliance becomes an inane waste of time. Like any activity mandated by others, regulatory requirements certainly can be scoffed at, put off, and rushed through like homework that children are so sure has no relevance to real life. But beware of the power of this self-fulfilling prophesy. If regulations *are* scoffed at and rushed through, with no expectation of results, benefits, or tangible gains, they assuredly will not have any relevance to your employees' lives.

A checklist approach is meaningless with performance-oriented Standards such as Hazard Communication and Lockout/Tagout because performance remains unaltered—nothing changes, no one learns. When no one learns anything, it just confirms what everyone *knew* to begin with: Compliance is a waste of time. "Why bother" becomes the prevailing attitude—*until* an inspector starts quizzing employees on the production line.

And isn't that the reverse of what it should be—of what actually makes a whole lot more sense? For whom are you complying and training, anyway? Nameless, faceless inspectors? Or the employees in whose individual lives and skills you have invested so much time and money? It is upon *them* that the future of your company depends, not the inspection of an OSHA or EPA official. It is *your employees,* not clean inspection reports, that either will carry you into the next century or leave you lagging in the dust of more progressive competitors. The question becomes not *Am I giving OSHA what they really want?* But, *Am I providing my employees with what they really need?* Comply for *them.* Train for *them.* Make your company safer for *them.*

If you want to comply with the intent of training regulations, which means changing attitudes and behaviors and enhancing awareness of what is safe and unsafe, healthy and unhealthy, then assess your compliance on the basis of the *quality of the results you achieve,* not on the presence of a training manual on a shelf, a box of tags in a drawer, or paper in a file.

How do you know if you're getting results? How can you tell if things are changing for the better in the workplace?

The Good Sense Factor

You set goals to be achieved through your safety and environmental compliance, and you monitor your progress toward them. As long as discernible progress is made, as long as things are changing, then it is a safe bet that you're complying.

In order to set meaningful goals, several things must be true:

1. You have to *know the regulations.* You have to understand the intent of the various safety and environmental regulations impacting your company (see Chapters 3-6).
2. You have to *know your workplace.* You can't develop safe operating procedures unless you know what happens on the production floor—from A to Z. You can't set goals to be reached through safe operating procedures unless you can recognize work hazards and unless you know the company's weaknesses and strengths—what it does efficiently and safely and where it is wasteful and unsafe.
3. You have to *know your employees.* You have to observe their behavior in the workplace. *Conduct inspections.* What do they do safely and well? What are their bad habits? Where do they tend to take shortcuts? *Talk to them.* Is there some aspect of the safe operating procedures that they don't understand? Are there any inconsistencies? Can they think of a better, safer way of doing a particular task? What would *they* like to see improve?

In evaluating procedures and setting goals for improving the workplace, don't hesitate to use the "Good Sense Factor." This means exactly what it sounds like it means. Does the procedure or "rule" the employee is following *make good sense?*

Sometimes, as a result of being separated from the production staff because of physical distance (their office is not at the production facility) or miscommunications, the people responsible for safety and environmental compliance will impose rules that just don't make any sense.

Case in Point: At a chemical manufacturing company, the Vice President and the Safety Manager established as company policy that workers in a certain chemical blending area were to wear safety goggles and respirators whenever they were breaking down and cleaning the equipment. The first time that a worker was found without either goggles or respirator, he or she would be given a verbal warning, the second time a written warning, and the third time would be sent home for three days. The Safety Manager passed this information on to the Plant Manager. The Plant Manager enforced this rule by browbeating his workers whenever he found them without a respirator *hanging around their neck*. As far as he was concerned, as long as they had the respirator on their body, it was fine; they didn't have to be breathing through it. But if he found them without one on more than twice, he'd fire them. Working with respirators hanging around their necks was cumbersome for the employees and caused the respirators to become damaged and dirty much sooner than it would have if they had been encouraged to wear them properly. Since wearing the respirators around their necks didn't make any sense to them, but they knew they would be fired if they didn't, the workers developed a very negative attitude toward the Safety Manager, the Plant Manager, and respirators in general.

The situation at this chemical manufacturing company impacts the company's compliance with the Hazard Communication Standard, the Personal Protective Equipment and Respiratory Protection Standards, and potentially, Lockout/Tagout, since it has to do with breaking down and cleaning equipment. The Safety Manager is in charge of an on-going program to comply with these regulations. In order to get a feel for how successful his compliance efforts are, what should the Safety Manager do?

First, he needs to develop one safe operating procedure for breaking down and cleaning the process equipment that includes any intersections with these regulations.

For instance, the procedure should include any steps to be taken to:

1. *Minimize energy hazards*, such as closing and locking valves or electrical controls and notifying other employees of the activities being conducted.
2. *Minimize physical hazards*, such as ventilating the area, eliminating ignition sources, cleaning up spills, and avoiding contact with incompatible materials.
3. *Minimize health hazards*, such as ventilating the area, minimizing skin contact and breathing of vapors, and wearing gloves, goggles, and respirator.

Second, he needs to train the workers in this procedure and talk to them about it. This dialog would serve to clear up the confusion about wearing the respirators around their necks. If the workers mention aspects of the procedure that they feel are inconsistent or unmanageable, he needs to change them.

Third, he needs to set a goal (with or without the knowledge of the workers) that *he* plans to achieve through this procedure. In this situation, his goal may be: "Getting the men to wear respirators properly while cleaning process equipment."

Fourth, he needs to monitor his progress toward that goal by randomly visiting and watching the men at work. He should invite the Plant Manager on his random inspections, but he cannot turn the whole matter over to him. It is very important that the Safety Manager himself sees whether or not his compliance efforts are working. If he sees workers not wearing the equipment properly, he needs to talk to them on the spot, perhaps even hold an informal group meeting to remind them of the procedure and ask them why they are not following it. He should try to stay away from threats of disciplinary action as long as possible, choosing instead to involve the workers in the refinement of the safe operating procedure to make it practical.

When, on several consecutive inspections, the Safety Manager witnesses the workers adhering to the safe operating procedure, he can consider the safe operating procedure, and thus his respiratory protection compliance efforts, a success. But his inspections, employee discussions, and openness to the revision of the safe operating procedure must continue in order to insure that it stays that way.

As discussed in Chapter 1, employees tend to know the most about their job and how to do it in the most efficient and safest manner. Far too often they remain an untapped resource regarding "a better way of doing things." A better way that often goes undone.

From the moment they are hired, production workers are told what, when, and how to do their jobs. Who thinks of *asking their opinion?* Who listens to what *they have to say*? Of course, many production workers would never dream of speaking up, even if asked. They would be afraid, because of a social or racial bias, that their suggestions would be turned against them, viewed as criticism of the boss and grounds for dismissal.

The double tragedy in this is that good ideas about production and safety left unexpressed and unacted upon can metamorphose over time into a staid unwillingness to change and a resentment of new procedures imposed by someone else. Given this seemingly no-win situation, what's an employer or well-meaning Safety Manager to do?

Whenever possible, base new procedures on employee suggestions and ideas.

This points out, one more time, the vital necessity of communication in compliance. Unless you talk to your employees, unless you walk around and observe them, unless you keep your finger on the pulse of their needs, you won't be serving their needs—which means you won't be complying.

PRIORITY PROCEDURES

Clearly, incorporating safety concerns into the operating procedures for each job title or job task at your company is one method for incorporating safety into daily production operations and achieving site-specific compliance goals. As a starting point, however, tackling this job might appear more than a little daunting.

So let's prioritize. How can you get the most regulatory coverage and the most results for your procedure? Are there any tasks at your company that are "everyone's job"? Are there any production responsibilities that touch every employee? Are there any specific duties that intersect with each employee's individual job activities, thus drawing them all together?

Generally, there is one category of tasks that touches everyone in your production department, regardless of their individual job titles or duties, and that is housekeeping. It just so happens that procedures developed for housekeeping can have a greater impact on your overall safety and environmental compliance than those developed for any individual job, if they are followed consistently. By *every employee.*

More than Pushing a Broom

What is housekeeping, anyway? Emptying garbage cans? Washing windows? Sweeping the floor? Isn't that what you pay a janitorial service to take care of?

Not exactly.

Housekeeping means maintaining the workplace in optimal operating condition—the *whole workplace.* Think of your workplace as if it were one giant machine requiring consistent lubrication, adjustment, tuning, cleaning, and rotation and replacement of parts in order to function properly. Viewed this way, maintenance of individual machines and equipment (as discussed in Chapter 6) is actually a *kind of housekeeping*—just one aspect of keeping the entire workplace at its most operationally effective. There are four others: storage of materials, control of inventory, disposal of waste materials, and control and clean-up of spills and leaks.

From fire prevention to waste minimization, there is no aspect of your regulatory compliance that isn't tied in with housekeeping procedures. In fact, the connection is so tight that it is virtually impossible to maintain compliance with any safety or environmental regulation unless your employee training emphasizes housekeeping. You say you don't really have any housekeeping procedures? Then at least two things are true and neither of them are good: you are not in compliance with the regulations impacting your company, regardless of what it looks like on paper; and you are wasting a lot of money.

So what do you do? You can't develop housekeeping procedures or talk about housekeeping in your training classes until you have assessed your company's housekeeping needs and you understand how those needs intersect with your regulatory compliance. The best way to accomplish this is to conduct a facility audit of the four areas of housekeeping listed above. This audit has the potential to be the greatest gift you can give yourself and your company. If approached objectively, with no motive other than the *discovery of what is really going on,* a housekeeping audit can teach you more than you would imagine, not only about the safety of your workplace, but about the *waste* occurring in your workplace in terms of raw materials, time, and, of course, dollars.

To help you get started, let's review what you need to be on the lookout for during your audit and see how what you find impacts your compliance.

Storage

What could be more basic to the manufacturing or service industry than storage? Everything you use to make a product or provide a service, all of your raw materials and all of your wastes must be stored somewhere, somehow, for some length of time. Regulations tell us that, in and of themselves, some materials are hazardous and others are non-hazardous. Just for a moment, forget about regulatory definitions and consider everything you store as being *potentially hazardous*.

Why? Because regulatory definitions will tend to put blinders on you, tempting you to concern yourself only with those materials that meet the regulatory definition of "hazardous," while you ignore the potential hazards of the multitude of materials that regulations overlook. The fact is that many materials defined by OSHA and EPA as "non-hazardous" or "non-regulated" from cardboard boxes to empty paint cans to solvent-soaked rags to waste lubricating oil, are hazardous *if stored improperly*.

In order to insure that safe materials storage is occurring in the workplace, you have to look at the workplace as a whole—not just at certain segments of it. Open your eyes. Don't limit yourself to what you think the inspectors are concerned with. Try to view your workplace as one interconnected unit, where everything has potential impact on everything else.

To make sure that drums of hazardous waste are stored properly, while ignoring the unsafe accumulation of regular combustible trash just several feet away, is self-defeating. To require the grounding of drums of flammable solvent in a chemical storage room, but to allow cans of flammable paint to remain uncovered at individual work stations, is illogical. To keep solvent rags in special fire cans, while tossing used paint filters into an outdoor dumpster, is inconsistent and dangerous.

When you conduct your audit, run through a list of questions for each material or waste you come across:

- Can it burn?
- Can it ignite spontaneously?
- If it is a liquid, does it evaporate easily?
- What can ignite it?
- If static electricity can ignite it, is it grounded?
- Is it stored near an ignition source?
- Is it under pressure?
- Can it explode when heated or dropped?
- Is it stored where it can be heated or dropped?
- Is it incompatible with anything stored near it?
- If it spills or leaks, will it run outdoors, onto soil, into water or a drain?
- Is it contained?

Storage and handling of the materials at your workplace are covered, directly or indirectly, by several safety and environmental regulations. Establishing one housekeeping procedure that addresses storage of all the materials in your workplace and then training all of your employees in that procedure will bring you a long way toward your compliance with the storage and handling requirements of each of these regulations:

1. *Hazardous materials management*—Storage and handling procedures for *hazardous materials* are included on Material Safety Data Sheets and are covered by the Hazard Communication Standard. (If, during your audit, questions arise regarding the hazards of a chemical you are storing, the Material Safety Data Sheet can serve as a helpful resource).
2. *Hazardous waste management*—Storage and handling procedures for *hazardous wastes* are included in RCRA's Hazardous Waste Generator Requirements, Title 40 CFR Part 264.
3. *Fire prevention*—Storage and handling procedures for *materials and equipment,* to insure that they don't cause or increase fires or obstruct exits, are covered by OSHA's Standard for Employee Emergency and Fire Prevention Plans.
4. *Pollution prevention*—Storage of *everything you may one day dispose of* is covered by pollution prevention and waste minimization regulations. These regulations emphasize the ways in which housekeeping procedures that include storage can be used to preserve the quality and extend the life of materials. So materials can be used, not disposed of. They also talk about how storage procedures can be used to avoid air pollution through evaporation, as well as soil and water pollution through spillage. When materials must be disposed of, proper storage of the waste to avoid contamination can make it possible for the material to be recycled and reused.

Storage of materials was mentioned throughout Chapters 3, 4, and 5. For most managers, it is not difficult to recognize, at least intellectually, the importance of materials storage in protecting the health of employees, limiting environmental impact, and preventing emergencies. You see the connection, but you are not necessarily motivated to get down to the brass tacks of changing storage procedures out on the production floor. Why? Because you don't realize that what the regulations call "proper storage" can save you money.

You're skeptical? That's okay. As part of your audit, go ahead and calculate how much money the company spends in salaries, paperwork, labor, and disposal costs getting rid of materials "wasted" through improper storage (see Figure 7-1).

Storage procedures impact every production employee. Hundreds of times each day your employees are faced with the opportunity to store materials properly or improperly. This means that several hundred times a day each employee has the

HOUSEKEEPING AUDIT #1: MATERIALS STORAGE

YES   NO

☐ ☐   Is combustible garbage stored in labeled trash barrels?

☐ ☐   Are trash barrels emptied regularly?

☐ ☐   Are combustibles, such as paper and cardboard, stored away from ignition sources, such as brazing and welding torches?

☐ ☐   Are all materials stored in such a way that aisleways and exits are not blocked?

☐ ☐   Are lids and bungs on containers (cans, jugs, drums) of hazardous materials; e.g. solvents, coatings, oils, fuels, cleaners, corrosives?

☐ ☐   Are flammable and combustible liquids stored away from ignition sources?

☐ ☐   Where they are handled and exposed to air, is ventiliaiton provided?

☐ ☐   Are incompatible hazardous materials stored separately from one another and from other materials with which they might react?

☐ ☐   Are drums of hazardous materials and hazardous wastes stored in stacks one container high, with aisle space between rows?

☐ ☐   Are lids or bungs on containers of hazardous waste?

☐ ☐   Are compressed gases stored upright and secured so that they can't be knocked over?

☐ ☐   If volatile solvents are pumped into or out of cans, drums, or tanks, are those containers grounded?

☐ ☐   Are hazardous materials (including aerosols) and hazardous wastes stored out of the sun and away from heat sources that might cause them to expand or explode?

☐ ☐   Are hazardous materials, hazardous wastes, and containers and equipment contaminated with either hazardous materials or wastes stored out of the rain where run-off might be created?

☐ ☐   Are hazardous waste streams kept separate from each other, and is non-hazardous trash kept out of them?

☐ ☐   Where run-off, overflow, leak, or spill is possible, is the storage area diked or otherwise contained to prevent the hazardous material or waste from reaching a drain, soil, or waterway?

☐ ☐   Are all raw materials stored in such a way that their usefulness or quality will not be reduced by exposure to heat, cold, water, humidity, or air?

Figure 7-1.   This Housekeeping Audit should be used as a Tool to Help you Systemize and Organize an Audit on Materials Storage.

opportunity either to enhance or detract from the safety of the workplace and the company's compliance status. When you can get this much mileage out of one standardized procedure, how can you afford not to make it a priority to develop safe operating procedures for materials storage and to train your employees to follow them?

Inventory Control
When you first think about it, inventory control might not seem to be a housekeeping concern. It seems more a management bailiwick, handled jointly by the Production Manager and purchasing department. It is. But inventory control, like materials storage, also impacts every employee who uses raw materials at your company.

The term "inventory control" simply refers to the regulation or monitoring of the use of the materials you store. The materials you have in inventory or in storage can be controlled in two ways: *at the time of purchase* and *at the time of use*. The degree to which the purchase and use of inventory is controlled or left uncontrolled will influence the ease or difficulty of establishing safe operating procedures for materials storage and of complying with the regulations that have storage requirements. *It's all connected.*

1. *Buy only what you need*—If procedures aren't established to the contrary, the temptation is always there to buy in bulk in order to get a discount price. More often than not, the trouble this creates costs more than the discount saves.

   When hazardous materials are purchased in volume, this can create logistical problems that make adhering to safe storage procedures difficult if not impossible. Where are you going to put the extra inventory? If you don't have enough space to store materials properly, what are you going to do? You're forced to cut corners, to deviate from storage procedures by storing materials in production areas or outdoors in the elements, rather than in designated storage areas. Because of space limitations, you may be forced to stack containers unsafely or to store incompatible materials together.

   Additionally, when you buy in bulk you risk never being able to use the materials at all. This can happen because the material's period of usefulness expires. If you have a storeroom full of expired paint, you are faced with the unhappy prospect of having to pay to dispose of it. Where are your savings when that happens? Or you might find yourself unable to use a backlog of raw materials because your process changes, someone higher up the line than you decides to switch to a new type of process chemical, or a product line is discontinued. If the manufacturer won't take the materials back, you're faced with the prospect, once again, of having to "waste" all of it.

   As you conduct your audit, make a list of products that appear to be overstocked and see if you can send them back to the manufacturer posthaste. The sooner you cut your inventory down to a manageable size, where raw materials are regularly turned over, the better for your company. In order to avoid these problems in the future, include in your Safe Operating Procedures for Housekeeping standards for limiting the purchase and therefore controlling the storage of all raw materials.

2. *Control accumulation of "nonstandard" materials*—If procedures aren't established to the contrary, nonstandard materials have a way of trickling into your workplace. Nonstandard materials include products that employees bring from home, products purchased for a one-time-only project, and samples brought into the workplace by salespeople.

   Failure to control accumulation of these products can lead to an array of problems:
   a. It can lead to overcrowding and thus improper storage practices, as mentioned above.
   b. It can lead to confusion and "mistaken identity" as employees struggle to distinguish between standard materials and others that just happen to be lying around.
   c. It can make compiling a Hazardous Materials Inventory in compliance with the Hazard Communication Standard a nightmare. That inventory is supposed to list *absolutely* every chemical product in your workplace. If you go by approved purchase orders to compile this inventory and you have taken no steps to control accumulation of nonstandard materials, you can be sure that there are at least *half-again as many* products out in your workplace that you are leaving off your inventory. Even though these products are nonstandard and mostly go unused, if you store them in the workplace, you are responsible for listing them on your inventory and obtaining a Material Safety Data Sheets for them. Wouldn't it be easier not to allow them to accumulate in the first place?
   d. Since they are nonstandard, these materials are seldom covered by any established training on hazardous materials storage and handling, such as Hazard Communication training. If employees are not trained in the hazards of these materials, it is unlikely that they will use proper protective equipment when handling them. They may be unaware of the characteristics of the materials, such as whether or not they are flammable or reactive, and thus handle the materials in a manner that endangers both themselves and the workplace in general. This liability can be avoided entirely by establishing procedures to monitor the accumulation of nonstandard materials.
   e. It can force you to spend a lot more money on waste disposal than you should have to spend. What happens to samples of coatings, wash-up solvents, adhesives, and cleaners brought into your workplace by salespeople? Or the materials purchased for a special order project, the client who had to have a special glue or paint? Occasionally, these products are used up. Far more frequently, they are used *once* and then stuck in a cabinet, on a shelf, on the

floor under a work table, or in the storage room, where they *stay,* accumulating dust until everyone forgets where they came from and what they are. And then the day comes when you have to dispose of them as hazardous waste. You can avoid this by informing salespeople that, if they leave sample materials, they will be responsible for picking up the remains of the sample within a certain time period unless you place an order for that material. You might get them to sign a statement to this effect whenever they call on you. As for special order jobs, care must be taken to buy only what is needed to complete that one job. The logic "Oh, might as well buy the larger container, we'll use it up eventually" is rarely sound.

As you conduct your audit, make a list of all the nonstandard materials you find, and then set out a plan to get rid of them. Remember, you can't keep house unless you have control over what you are storing; you can't operate safely, nor can you minimize waste, unless you control the type and quantity of the materials you store. So establish Safe Operating Procedures for Housekeeping to help you gain this control.

3. *Use raw materials efficiently*—If procedures aren't established to the contrary, inefficient methods of using raw materials will sneak into your employees' ways of doing things. This statement isn't meant as criticism of your employees or as a suggestion that they are trying to sabotage the workplace. It's simply fact. Using raw materials inefficiently often *seems* easier and faster.

   As you conduct your audit, check on the following:

   a. Are employees *using raw materials sparingly and in correct proportions*? For example, are they mixing chemicals according to manufacturer's recommendations, or according to their own beliefs about what works best? When wash-up solvents are required, do they try to make a little go a long way, or are they careless? Do they reuse a degreaser as many times as possible, or do they dump it into a waste barrel after the first use? Do safe operating procedures need to be established to control the use of raw materials?

   b. Is equipment *adjusted properly so that materials aren't overused or wasted*? For example, is an excess of mold release being sprayed into your foam seat molds because the controls are adjusted wrong? Is more than enough glue being applied by your bindery machine? Are the nails that pass through your electroplater being coated with too much zinc? Is overspray from your powder paint line piling up on the floor? Sometimes, using too much of a lubricating fluid will make a press run faster, but also less efficiently, because it creates more waste and vapors. Do safe operating procedures need to be established to monitor the control of material application?

   c. Are materials *properly diluted*? For example, cleaning solutions are the worst offenders when it comes to tempting employees to use them full strength. When purchased for industrial use, most floor cleaners, car washes, tire cleaners, engine shampoos, toilet cleaners, laundry detergents, stain removers, sanitizers, glass cleaners, air fresheners, and all-purpose cleaners have to be diluted with water to be most effective. Often, if used full strength these products will damage whatever they are supposed to clean.

Initially, you might not think of these production concerns as having much to do with inventory control. In fact, monitoring inventory is your best indicator of whether or not raw materials are being used efficiently and is one important indicator of whether or not they are used safely.

If revamping your inventory controls looks to be about as inviting as changing your storage procedures, turn your attention for a moment to a calculation of the cost of managing and disposing of excess inventory and nonstandard inventory. Can you put a dollar figure on the liability associated with improper storage necessitated by excess inventory? What would it cost you if you if an employee got hurt as a result of using a nonstandard product that he or she wasn't trained in? And what do you think your company might be spending in inventory that is lost through inefficient usage, and only ends up on the floor or in the air, causing additional waste and pollution problems? As part of your audit, try to estimate what your inventory costs you.

Waste Management

How do you manage your waste? Not just hazardous waste, but *all the waste* your company creates.

Storing waste is certainly an aspect of managing it properly. The importance of monitoring the storage of waste materials was mentioned earlier. In order to truly *manage* your waste, however, you have to control it *before it is stored*. Sometimes, once waste is stored, it becomes part of the landscape of the workplace. Daunted by the cost or inconvenience of removing and disposing of the waste, management allows it to remain inside the workplace or outside, tucked away somewhere on the facility grounds. Employees learn to work around it. Time passes. And sometimes, if enough time goes by, it is forgotten entirely—until it is the source of *a problem*. What kind of problem? Fire, injury, soil contamination, run-off, water contamination, mosquitoes, and rodents.

In order to manage your waste, you have to focus on the waste *as it is accumulated* and put into effect procedures to control that accumulation. What kind of waste do you need to manage? All kinds.

1. *Hazardous waste*—With respect to your hazardous waste management, controls regarding its accumulation, such as where and how long it can be stored and how it must be labeled, are imposed upon you by RCRA Title 40 CFR Part 262. In order increase the odds that these controls are being followed in the workplace, they need to be incorporated into the Safe Operating Procedures for

Housekeeping, not held at arm's length as the sole responsibility of one or two people. Everyone needs to know about them. And it needs to be everyone's responsibility to make sure they are met.

2. *Waste piles*—Any accumulation of waste outside your building, on your property, and exposed to the elements can be considered a waste pile. There was a time when companies could use their grounds as a personal dump site. Who cared, right? You could pile up old equipment, broken down vehicles, out-of-date raw materials, empty drums and other chemical containers, broken appliances, wood, old roofing materials, cardboard boxes, tires, batteries—anything you wanted out of the workplace but didn't want to go to the trouble to haul away. If the broken equipment leaked oil, nobody cared. If the old batteries leached lead into the soil or the tires bred mosquitoes, it was nobody's business but yours. If rainwater washed ink off of discarded printed matter, nobody noticed.

   Those days are over. Thanks to pollution prevention legislation, your waste is no longer your own. You have to be able to account for its impact on the surrounding environment—on air, soil and water. But, instead of fretting over that, why not abolish waste piles entirely? That's what proposed waste minimization regulations say you should do. These regulations say that, through standard operating procedures, you should be able to control accumulation to such an extent that piling up excess waste will become a thing of the distant past. How do you do this? By controlling inventory as mentioned above, by returning unused products to their manufacturer, by reclaiming and reutilizing raw materials and equipment, and by establishing safe operating procedures that encourage innovative thinking on the part of employees and provide a forum for experimenting with waste reduction ideas. By controlling waste before it accumulates, you will reduce your waste disposal and clean-up costs, reduce your environmental impact, and create a safer, healthier workplace.

3. *"Empty" containers*—Empty containers are one of the most overlooked hazards of the workplace. In order to understand this, you must realize that empty containers aren't empty just because there is no more usable material in them. Unless they have been triple-rinsed, empty containers hold a residue of the original contents that *retains the same hazards* of the original contents. Sometimes this residue can be even more hazardous than if the container were full. For instance, if an "empty" can of thinner, gasoline, or some other volatile solvent is closed and then left in the sun, the trunk of a car, or other hot environment, the residual vapors in the can will expand as they are heated, resulting in the explosion of the can. Pressurized containers, such as aerosols or compressed gases, are as hazardous when empty of usable product as they were when full. When heated or punctured, "empty" aerosols will still explode. If the valve end is knocked off an "empty" compressed gas canister, it will still become a rocket. Obviously, the accumulation of "empty" containers of hazardous materials can be dangerous,

resulting in fires, injuries, and other emergencies. Safe operating procedures need to be established to insure that accumulation is controlled.

Awareness of the non-emptiness of seemingly empty containers also plays a role in preventing the mixing of incompatible materials and wastes. For instance, an "empty" spray bottle that once contained an alkaline detergent should not be used to hold an acidic wire cleaner. Or an "empty" drum that held a corrosive liquid should not be used, unless rinsed first, to contain a flammable liquid waste. And since empty containers are not empty, they should never be left out in the elements, where they can fill up and overflow with rainwater, contaminating the ground with residual product set in flight on wings of water. Procedures governing the management of empty containers are a crucial part of effective housekeeping.

4. *Solvent wipes and paint filters*—These waste materials demand special consideration because, depending upon how they are managed, they can be *spontaneously combustible*. When these materials are stored in a confined space, where the circulation of air is poor, the rag and oil or filter medium and paint may undergo a slow oxidation. Under these conditions, the material retains the heat of combustion, which can raise the temperature of the material to its autoignition temperature. When this slow accumulation of heat results in burning, it is called spontaneous combustion.

Solvent wipes piled in the corner of a poorly ventilated storage room or paint filters in an outdoor dumpster are prone to undergo spontaneous combustion. Each year, millions of dollars in property are lost because of fires that originate in this way—all of which could have been prevented through safe housekeeping procedures.

As you conduct your audit, check and see how you're doing in your management of all of these waste materials (see Figure 7-2).

Spills and Leaks

OSHA and RCRA may have conditioned you to think of spills and leaks as "special circumstances," uncommon occurrences requiring unusual response. Sometimes this is true. Far more often, however, spills and leaks are small, subtle, nearly unnoticeable daily events that result in problems only cumulatively, when added up over time.

Such spills or leaks occur for several reasons, all of which can be controlled by your employees as they conduct their regular work. Be on the lookout for these and other sources of leaks and spills as you conduct your audit:

1. Leaking valves on chemical storage containers and dispensers (jugs, drums, totes, tanks).
2. Leaking gaskets on pumps.
3. Punctures or tears in materials transfer hoses.

HOUSEKEEPING AUDIT #2: WASTE MANAGEMENT

YES NO

☐ ☐ Is hazardous waste being accumulated only in designated areas?

☐ ☐ Are all containers of hazardous waste properly labeled?

☐ ☐ Are all hazardous waste streams segregated from each other and from non-hazardous wastes?

☐ ☐ Is all hazardous waste removed from the property and disposed of in the appropriate time frame?

☐ ☐ Are there no accumualtions of out-of-date or unusable product (in the basement, the boiler room, or detached storage building, etc.) that need to be disposed of?

☐ ☐ Are containers of hazardous materials or hazardous wastes (whether "empty" or full) accumulated only in areas that are protected from the weather (sun, rain, freezing temperatures)?

☐ ☐ Are "empty" containers of hazardous materials handled in such a way that their hazards are minimized?

☐ ☐ Are "empty" containers of hazardous materials triple rinsed before they are used to contain a *different* material or waste?

☐ ☐ Are accumulations of solvent wipes or rags or paint filters monitored to prevent spontaneous combustion?

☐ ☐ Are waste piles of any kind removed from the company property?

☐ ☐ Have old machines, equipment parts, vehicles or appliances been removed from the company property and disposed of or recycled?

☐ ☐ If any of them were leaking lubricating fluids, refrigerants, or fuels, have they been cleaned up?

☐ ☐ Have all accumualtions of tires, batteries, scrap rubber or metal, wood, cardboard, or paper been removed from the property and disposed of or recycled?

☐ ☐ Is no part of the property used as a "dump?"

Figure 7-2. This Housekeeping Audit should be used as a Tool to Help you Systemize and Organize an Audit on Waste Management.

160   Safety and Environmental Training

4. Failure to use drip buckets during materials transfer (or empty them afterward).
5. Failure to drain hoses after use.
6. Failure to transfer materials in a contained area.

If uncontrolled, these kinds of routine spills and leaks can have several negative impacts on your workplace. Leaks due to the continued operation of broken or faulty equipment are *expensive*. So are spills that occur because of employee failure to take simple preventive measures. They cause you to waste raw materials that end up on the ground. You have to pay your employees to clean up the mess, and then you have to pay to dispose of it.

Leaks and spills create an *unsafe* work environment. When materials are sprayed in the air due to faulty valves or gaskets, exposure hazards are created. Materials that end up on the floor create slippery walking surfaces. Vapors from chemical leaks cause breathing hazards and fire hazards.

Safe Operating Procedures for Housekeeping must include proactive measures for preventing spills and leaks. Failure to include them will reduce the power of your housekeeping procedures to maintain a safe and compliant workplace by at least 25%. Employees must understand that, regardless of what was "standard" in the past, they are not expected to *make do* with leaking parts or faulty equipment any longer. Neither are they expected to cut procedural corners in an effort to save time. In fact, it is essential that they know that these expectations *have been reversed.*

Clearly, housekeeping procedures that include storage, inventory control, waste management, and spill/leak control can have a profound influence on the well-being or wellness of your company. Housekeeping impacts safety, compliance with nearly every safety and environmental regulation, and, in a very real yet often overlooked manner, the budget. But in order for this influence to be realized, housekeeping must be included in the day-in, day-out job description of every employee. As you conduct your audit, ask yourself this: "How can I make "keeping house" part of the work ethic at this company and not something everyone jumps to do because visitors are coming?"

If your goal is to develop safe operating procedures that are relevant to your company and your compliance, you can't do it from behind your desk. Whether you are developing a safe operating procedure for one job or for housekeeping plant-wide, you have to pick up a copy of the regulations, a pad and pencil, walk out onto the production floor, and start watching what goes on, asking questions, and listening. Begin with one department, one line of chemicals, or even one machine. Discover everything you can about how your employees do their jobs. Investigate the materials your employees use, the shortcuts they take, the problems they encounter, the sensations they feel, and the reasons they operate as they do. Then build a safe operating procedure around the needs that they reveal. Don't leave well enough alone. Focus on renovations. Find out what's wrong now.

And before it breaks, fix it!

8

Combine and Conquer

Man shapes himself through decisions
That shape his environment.

René Dubos

No matter how much you might want to, you can't do it alone. Safety and environmental training must be combined three ways to be conquered:

1. Safety training and environmental training must be combined with each other and viewed as part and parcel of the same subject with the same purpose.
2. Safety and environmental training compliance must be combined with with the responsibilities of every manager in the company, whether Safety Manager, Personnel Manager, Production Manager, or Business Manager.
3. Safety and environmental training compliance must be combined with the company's efforts to meet *all* of its environmental compliance obligations.

THE MANAGEMENT CONNECTION

We live in a society where specialization is revered to the point that jobs are isolated from each other by seemingly impenetrable walls of credentials, lingo, bureaucracy, and paperwork. It seems that we would almost rather *be* an expert, no matter how impractical or unintelligible, than have the ability to *utilize or benefit from our expertise*. The fact is, however, that if you want to conquer your safety and environmental training compliance and put it to work for you, you must knock down the walls that keep the specialists within your company from working toward a common goal.

The traditional division of labor: Safety Managers handle accidents and OSHA while Environmental Managers handle hazardous waste and EPA compliance. If you work at a company where these jobs are held by two or more people, you probably don't work together. You may not even see or speak to one another. Since both of you are involved in regulatory compliance, it seems natural that you would confer, coordinate wherever possible, and screen your policies, procedures, and training programs for contradiction or duplication. But this doesn't often occur. Instead, it's more likely that you don't know what the other person is up to and, perhaps, don't much care.

Since job titles vary from company to company, and since they don't necessarily match up with job responsibilities, they are not what is important. What *is* important and relevant to the task of combining safety and environmental training with other compliance activities is identifying the individuals within your organization who have been given responsibility for *any part of it.*

Are you one of these individuals? Then take a moment to examine how you do your job: Do you use your specialty, your expertise, as an opportunity to wield arbitrary power? Are other managers or employees scared of you? Do other managers and employees have no idea what you, as the Safety or Environmental Manager, really do? Or do you make it a point to communicate with other people within your organization who have the responsibility for safety and environmental compliance, or who are impacted by it? Do you know what the other Safety or Environmental Managers within your company do? Are you trying to develop a cohesive safety and environmental program, or are you trying to protect your turf? Do you make an effort to combine safety and environmental training and compliance needs with the needs, procedures, and policies of other managers? Do you let other managers know what's going on? Do you ask for input?

Case in Point: At a midwestern treatment, storage, and disposal facility, Dave, the Environmental Compliance Officer, is responsible for maintaining the facility's compliance with its Part B Permit. Once each day he conducts and documents a plant inspection, as required by the Permit. The inspection involves drum storage, handling, and labeling, tank farm operation and hose management, spills, and ground contamination. Dave files the inspections in a looseleaf notebook that he keeps at his desk, ready to show the state or federal inspectors at their next visit. At this facility Jerry, the Purchasing Manager, buys safety equipment for the plant and is also responsible for insuring the plant's OSHA compliance. Once a week Jerry conducts a safety inspection. He checks things like general housekeeping, fire hazards, drum management, labeling, tanker loading and unloading, use of respirators and other safety equipment, spill clean-up, and ground contamination. Jerry, like the Environmental Compliance Officer, files his inspections in a notebook that he keeps in his office. Occasionally, both Dave and Jerry will send a memo to the Vice President, letting him know the results of their inspections. But they never talk about their findings with each other. They never discuss what might be done to remedy recurrent problems. It has never occurred to them to sit down and put their heads together.

The problem here is that, first and foremost, permit compliance and safety are seen as two separate, unrelated jobs. No activity conducted to comply with environmental regulations, from inspections to training to equipment modifications, is seen as intersecting with safety compliance. And nothing done in the name of safety or OSHA compliance is seen as having anything whatsoever to do with complying with an EPA permit.

Secondly, at this facility, the Environmental Compliance Officer has more clout than the Purchasing Manager. The Environmental Compliance Officer has more education and credentials and makes more money than the Purchasing Manager. He sees no need to "stoop" to safety concerns when he's been educated and trained to be an expert in environmental compliance. The Purchasing Manager, by contrast, has no formal safety training. He has taught himself virtually everything he knows about safety, and frankly, he's proud of it. He feels that since he's not just the Purchasing Manager anymore, but the Purchasing/Safety Manager, he's more valuable to the company and has more job security. He sees no need to "give away" his hard earned knowledge and position to the Environmental Compliance Officer. In Jerry's mind he has nothing to gain and everything to lose by combining his expertise with Dave's.

Although the specifics of title, training, and responsibility vary from company to company, the bottom line is this: *the issues of clout, status, and job security often come between those of us who are responsible for safety and environmental training and compliance and our ability to meet our compliance obligations effectively and efficiently.*

If, as a Safety or Environmental Manager, you are *not* motivated by self-interest to keep your job distinct from everyone elses, could it be that you simply have never examined the regulations impacting your company to see where safety management and environmental management intersect? Perhaps combining safety and environmental training and coordinating safety and environmental management has never occurred to you. You may even think that it's not allowed.

It seems likely that this is true because even at companies where *one person* is responsible for all safety and environmental training and compliance, that person will tend to handle them separately and draw no connection between them.

The traditional division of labor: Production Managers are concerned with quotas, schedules, product quality, and on-the-job training; Business Managers care about payroll, taxes, budgets, profit, and loss; and Human Resources Managers worry about hiring, insurance, employee complaints, and orientation. People who hold one of these jobs are directly impacted by safety and environmental training. But most don't know it. They do not see the connection or chose to ignore it, believing it insignificant. Why do they feel this way? Perhaps because regulatory compliance isn't their specialty and they're afraid of it. Perhaps because they don't know enough about what goes on beyond their management niche to recognize their dependence on efficient, effective safety and environmental compliance. Or perhaps because everything they know of regulations *is negative.*

Is it possible that, as far as they're concerned, safety and environmental compliance is impractical because it has nothing to do with production, money management, hiring, and firing? Perhaps they believe it has nothing to do with running a business well. They view it as busywork created by government bureaucracy. They think the less time, money, thought, and effort that is wasted on it, the better. Perhaps they see only negatives and are blind to the benefits of combining forces with the Safety and Environmental Managers at the company.

Could it be that instead of making safety and environmental training compliance their business and integrating it into their management style, decisions, actions, and priorities, the Production, Business and Human Resources Managers fight against it and whoever is responsible for it? Instead of learning about safety and environmental training and compliance and figuring out where it intersects with their particular area of management expertise and how they might be able *to use it to their advantage,* they do their best to pretend it doesn't exist. Since they know of nothing to be gained through compliance (at least nothing that will enhance their position or make them look good), they probably consider it bad business to invest more than the minimum in thought, effort, time, and money. Consistency, efficiency, and effectiveness in training and compliance is not an issue; *getting it over with* is.

Ring a bell?

When managers schooled in other specialties find themselves personally responsible for safety and environmental compliance, with these new duties added onto their job description, the tendency may be to view it at arm's length. To view it as an external, superfluous task. It's "not really my job," they may be tempted to think. "It's just something *extra* the government makes me do." So they don't look at it as part of managing production, managing personnel, or managing the company finances. Perhaps more significantly, they don't recognize the benefits of integrating it into the way they manage other aspects of the company.

The bottom line: When management *of the company* is not combined with management *of safety and environmental compliance,* managers often end up *working against themselves.*

Broaden Your Perspective

It may not appeal to you as a specialist, expert, or business person, but the days when you can just "do your job," whether that means operate a production facility, turn a profit, cut inventory costs, balance the budget, hire employees, file insurance, dispose of hazardous waste, write an air pollution permit, train employees to do a job, or change a production process, without considering how your decisions or actions *impact or are impacted* by safety and environmental training and compliance, are over. Forever. Safety and environmental regulations demand that managers broaden their roles.

In order to conquer and not be conquered by the regulations, as a manager you must alter the duties and responsibilities by which you define yourself. The irony of this is that broadening your role as manager rather than strengthening the walls around your self-protective niche is actually the best method available to you for securing your future. By combining forces with other managers at your company, you help guarantee that each person's decisions, actions, policies, and procedures bring everyone closer to fulfillment of the company's safety and environmental responsibilities. In doing so you limit the opportunity for counter-productivity, contradiction, duplication, and inconsistency in the leadership and management of the company.

This is not to imply that everyone in your organization should cater to the Safety and Environmental Managers. It is to suggest, however, the use of a system of checks and balances, where every manager holds every other manager accountable for the safety of the employees and the environmental integrity of the company as a whole. No manager is exempt from responsibility. No one can say it's not his or her job.

In order to do this, of course, meeting safety needs and operating the company in an environmentally healthy manner must become integral to each manager's *reason for being,* inseparable from his or her duties as Production Manager, Business Manager, Personnel Manager, Safety Manager, or Environmental Manager. It must be common knowledge that any policy, procedure, equipment, process, or training class that carries the company closer to fulfilling its safety and environmental obligations is in the *common good* of everyone at the company. It is also in the *individual interest* of every manager who wants to do a good job today and wants to feel secure in having a job tomorrow.

Case in Point: At a furniture manufacturing company, the Vice President is told by the corporate attorney that the plant employees are way behind schedule in receiving OSHA training. Right-to-Know classes haven't been conducted in almost two years. And, the attorney tells the V.P., respirator training promises to be a real bugaboo. The V.P. only has a peripheral knowledge of these regulations—one has to do with chemicals, the other with a protective mask. Both require training. It's mainly a paperwork problem. How complicated can it be? Still, he needs someone to take care of it. He's been having problems justifying the Human Resources Manager; the guy just hasn't been pulling his weight. The V.P. decides this OSHA problem just may be the answer. He calls a meeting with the Human Resources Manager to give him the job. When the men sit down, the V.P. tells the Human Resources Manager about his call from the corporate attorney's office. He tells him what they need is documentation of training, proof in a file before some inspector shows up or an employee files a complaint, that they've provided the plant employees whatever information the law says they're entitled to. He tells him to get some videos and books and do it himself or hire outside help—it doesn't matter which—as long as documentation of training is in each employee file very soon. He tells him he would have given the job to the Safety Manager or the Environmental Compliance Officer, but they're too busy with other projects. "You're the only one with the time to do it," the V.P. tells him, "so you'll have to take care of it by yourself."

Before we hear the rest of the story, let's look at what the Vice President might have done differently.

What might have happened if the Vice President had asked the attorney to send him copies of the specific regulations in question and then read the regulations himself? What would have been the advantages to the company if, after reading the relevant regulations, the V.P. had called a meeting of *every* manager whose cooperation or participation would be necessary to meet the requirements of the regulations? What would have happened if he had distributed a copy of the regulations to each manager before the meeting and then, when they were all together, asked them each specific questions about their interpretation of the regulations? What if he had asked for their suggestions regarding how to meet certain requirements of the regulations most effectively, or for their opinions regarding the value of compliance with the regulations to their particular area of the operation and to the company as a whole? If he had initiated such discussion, is it likely that any compliance effort undertaken by one or all of the managers present would have been successful not only in terms of *complying with the law,* but also in terms of *having tangible value for the company?*

The answer is *yes.*

Especially if the V.P., in opening the meeting, had said something like: "Tom is our Safety Manager and Nancy is our Environmental Compliance Officer, but safety and environmental compliance is part of *each of our jobs* as managers of this company. We have some training requirements that we've fallen behind in meeting. I'm not here to point fingers or lay blame, but to ask each one of you to work together to make this training a success, not only in terms of complying with the law, but because it makes this company a better place to work."

After this introduction the V.P. still could have given the job to the Human Resources Manager, since he would also have given him a network of support. It would have worked if he had said: "Since Bill has the freest schedule, he's going to take the lead in getting the training completed. Tom and Nancy are going to provide him guidance in exactly who needs which training and will help him decide who will conduct the classes, whether we need an outside consultant, and so forth. But I want each of you to feel you have a voice in this matter. It won't work unless we're all involved."

Unfortunately, however, the V.P. didn't give the Human Resources Manager a network of support. And so things didn't turn out very well for him, the employees needing training, or the furniture company.

Case in Point: The Human Resources Manager doesn't know whether to laugh or cry as he walks back to the office. He thought for sure he was going to be fired, but instead he's been given a job he knows absolutely nothing about and told to do it in 30 days. He has only been to the plant one time, on a tour the day he was hired. He's never met the Production Manager in person and has only dealt with the Safety Manager once. At his last job at a hardwood flooring company, he attended a couple Right-to-Know classes, so he's not too worried about those. But he has no idea which employees out in the plant need training. Every time he calls the Production Manager to get the numbers of employees in each department who work with chemicals, the guy puts him on hold forever, or tells him he'll call back with the information and never does.

Finally, after two weeks of this, the Production Manager blows up at him, saying: "I'm in charge of production, right? But when do we ever *produce* anything around here? All I do is shuttle my people from a lifting class to a defensive driving class to a fire extinguisher class. And now you want them for a chemical class. There was a time when people went to work to work, not to attend classes!"

When the Production Manager refuses to give the Human Resources Manager any information, the Human Resources Manager goes back to the V.P., who goes to the Production Manager, who eventually gets back with the Human Resources Manager. The V.P. and the Production Manager are in agreement that 20 minutes during lunch will cover the bases just fine.

So the Human Resources Manager shows a 15-minute video at lunch, gives the employees a standardized test, and puts a copy of each test in his files. The respirator training is put on hold until production slows down. The Safety Manager never knows that the training has been held.

The story of the Human Resources Manager illustrates what can happen when managers do not share safety and environmental compliance as a common goal. But the results can be just as ineffectual, and even dangerous, when various managers within a company have been given very specific safety and environmental compliance tasks, but have no real understanding of the *purpose* of their tasks or how they fit into the *overall compliance* of the company.

This happens because, for one reason or another, managers don't communicate among one another and share information. Communication is blocked either because managers are too busy protecting their small turf to share information or because most simply lack knowledge of how and why the organization is regulated.

Maximize Your Effectiveness

Case in Point: At the treatment, storage, and disposal facility mentioned early in this chapter where the Environmental Compliance Officer handles EPA Part B Permit inspections and the Purchasing Manager conducts safety inspections, the

Plant Manager conducts safety meetings with all plant employees, while the Director of Transportation conducts safety meetings with the drivers, who must load and unload trucks and tankers at the plant. This division of responsibility might work except for a couple of problems.

First, the Plant Manager and the Director of Transportation *never see* the inspections conducted by the Compliance Officer and the Safety Manager. Since the Plant Manager is only aware of the facility's EPA Permit obligations in a general sense, it is difficult, if not impossible, for him to talk about the things that plant employees are or are not doing to meet these obligations. His knowledge of specific OSHA regulations impacting the facility is even sketchier. So, instead of conducting weekly safety meetings with the guidance of the other manager's inspections and expertise, he conducts them by feel, based solely on his own common sense and experience. His meetings are fine, but certainly not nearly as effective as they could be.

By the same token, the Director of Transportation holds his meetings based solely on Department of Transportation (DOT) requirements. He never visits the plant while his drivers are loading and unloading; he doesn't have the benefit of the Safety Manager's inspections or the Plant Manager's observations. Consequently, his meetings are basically dry rehashings of the regulations, having little to do with the reality of the driver's jobs or the hazards they face.

The second problem facing these managers is the fact that none of them understands the big picture. They have been told to do a job and they do it. But they know so little about the regulations impacting their facility that they actually have no way of judging whether or not they are doing their jobs correctly.

Each of the four managers at the treatment, storage, and disposal facility, although directly involved in safety and environmental compliance, does his job in isolation, never thinking about how it intersects with another manager's job or how the company's overall compliance might be improved if they combined their efforts and shared expertise and information. What the managers lack is not care or interest in safety and environmental compliance, so much as *understanding* of their individual roles *not as ends* in and of themselves, but *as means to an end* that demands the combination of each of their individual efforts.

At the furniture manufacturing company, compliance has not been integrated into the management role, the management job description. There, compliance is viewed as an external, but necessary, evil—necessary only to the extent that it keeps the inspectors and lawsuits away. At the treatment, storage, and disposal facility, compliance has been combined with the role of the manager to a *limited* extent. Managers are involved in compliance, but it's a piecemeal, hunt and peck kind of involvement, without purpose. They aren't coordinated toward a common goal.

Case in Point: Bob, the Production Manager at a commercial printing company, says: "I remember when all I had to do was run a print shop, produce a quality product. There was a minimum of hand-holding you had to do. A minimum of babysitting. People came to work because they wanted to make a living. Now it seems they come to work for different reasons, because they want to get something from the boss by proving him wrong. They want to catch me with my pants down. The whole motivation for working has somewhat changed. And all these regulatory requirements, all these rights that belong to the employee, what have they done? Who've they helped? They've given new impetus to the employee to sit around and wait for handouts…and, as for me, they've made my job a damn sight more complicated. There was a time when I was a printer. A good printer. Now, ha! Now I'm the company expert in regulatory compliance. I spend my time holding training classes, documenting training classes, managing hazardous waste, labeling everything from 55-gallon barrels to circuit breakers, adding up air emissions, collecting data sheets, filling out accident reports, compiling chemical lists, checking respirators, leading fire drills, completing, copying, filing, and mailing endless numbers of forms. If it has to do with safety or EPA, then I have to do it or make sure it gets done. Do I like it? Hell, no. And I don't understand it either. I can't point to one improvement all this complying has made. I'd complain to someone about it, but I'm too busy trying to keep my head above water. And when it's all said and done, I have no choice."

Bob has had to change his job description to include safety and environmental compliance. But now it's controlling him. Compliance is running his business, wearing him out. There is a palpable sense of futility in his random listing of compliance obligations. They are disconnected. They're just items on a list of things he has to get done, but from which he doesn't know how and doesn't expect to benefit.

What could Bob do to conquer his compliance?

Mapping the Intersections
Bob is overwhelmed by the sheer numbers of regulations he has to comply with. They seem arbitrary and disjointed. There is no rhyme or reason to them. And why should there be? Bob is missing two key pieces of information about his safety and environmental compliance obligations that prevent them from making any sense.

1. He doesn't understand the intersections among the regulations themselves—where they overlap and how they can be combined.
2. He doesn't understand the intersections between regulatory compliance and the operation of his business. How he can use compliance, not only to enhance certain aspects of production and employee performance, but to improve the quality of the work environment.

Bob must understand this fact: *Regardless of the number of people responsible for regulatory compliance at a company, whether it is 1 or 20, a common goal or*

170   Safety and Environmental Training

goals must connect all compliance activities. If you are one person alone, like Bob, this goal will be the yardstick by which you can measure whether or not your compliance efforts are benefitting you, your employees, or the company. It will also help you keep your sanity, because it will enable you to feel as though you are working *toward something tangible,* rather than having your energies splintered off in multiple directions. If you are one of several people, as at the furniture company or the treatment, storage, and disposal facility mentioned earlier, a common goal or goals to be achieved through compliance is the only thing that will insure that you don't spend much of your time, at the very least, duplicating each other's efforts and reinventing the wheel, or, with more disastrous results, contradicting each other—*actually canceling each other out.*

But, back to Bob, let's look at the key safety and environmental regulations impacting his printing company and find the intersections.

Safety	Environmental
Right-to-Know	RCRA Hazardous Waste Management
Protective Equipment	RCRA Preparedness/Prevention and Contingency Plan
Respiratory Protection	Waste Minimization
Lockout/Tagout	
Emergency Action/Fire Prevention	

Your tendency, when you look at all of these separate regulations, may be to get caught up in the busywork they require. You may, like Bob, become distracted by the forms, documents, and plans, the special labels and equipment, all of which you must gather and sort, punch holes in, put in notebooks, post on walls, or stick on drums, copy, file, or mail. When viewed this way, the regulations seem utterly arbitrary—fancy scavenger hunts designed by bureaucrats who know nothing about industry. If you're lucky enough to gather all the pieces together by the time one of them knocks on the door, then you can say you've won the game.

Whoopee.

It's easy to see how anybody could get lost in all of the stuff these regulations require:

Right-to-Know
　　Chemical inventory
　　Material Safety Data Sheets
　　Labeling
　　Written Program
　　Training
Personal Protective Equipment
　　Gloves
　　Glasses

    Goggles
    Face Shields
    Clothing
    Inspections
    Training
Respiratory Protection
    Respirators
    Written Program
    Inspections
    Training
Lockout/Tagout
    Tags and locks
    Written Program
    Inspections
    Training
Emergency Action/Fire Prevention
    Written Program
    Evacuation Plan
    Inspections
    Training
RCRA Hazardous Waste Management
    Manifesting
    Labeling
    Annual Reports
    Inspections
    Recordkeeping
    Training
RCRA Preparedness/Prevention and Contingency Plan
    Preparedness/Prevention Plan
    Contingency Plan
    Evacuation Route Maps
    Personal Protective Equipment
    Inspections
    Training
Waste Minimization
    Written Plan
    Waste Tracking System
    Training

Viewed as eight unrelated "To Do" lists, these regulations have nothing in common. However, as established in Chapters 3-6, each of these regulations actually exists as a guideline for employers to use in influencing their employees' behavior in two areas: *Workplace safety, and the impact of the workplace on the environment.* Each regulation or guideline provides employers with a recommended system for changing employee behavior through, first, information, and secondly, the necessary policies, procedures, resources, and equipment to back up that information.

It follows, then, that providing information that will enable employees to work safer and minimize the impact of their actions on the environment must be the *first goal of each compliance effort* and the *common goal that connects every compliance effort.*

Training is the common thread.

Too often, however, this gets reversed. First and foremost, the busywork, the policies, procedures, resources, and equipment that the regulations require is what is emphasized. When training is conducted, it is almost as an afterthought; the last thing checked off on a list of compliance requirements.

Instead, training should be your *reason for doing the busywork.*

Enhancing the transfer of knowledge to your employees should be what you design all of the paperwork, programs, labeling systems, and equipment specifically to do. In fact, if they don't enhance your training by helping to influence your employees' behavior, then none of it is serving its intended purpose. And, therefore, *it isn't worth doing it at all.*

COMBINING TRAINING

Once you have a grasp of the purpose of the regulations impacting your workplace, you're just one small step from streamlining your compliance and making it more effective, by covering multiple regulations at each training class. All you have to do is identify the intersections among them. There are three main levels of intersection among the regulations: *content, personal protective equipment use, and attitudes.* Keeping the regulations covered in Chapters 3-6 in mind, let's find the intersections.

Content

These are the easiest to find. All you have to do is list the topics covered by several of the regulations requiring training and then identify where those topics overlap. Whenever a content overlap occurs between two or more regulations, you can conduct *one* training session that meets the training requirement for *every other regulation with that content.* Yes, it really is that simple.

For example, the *Hazard Communication Standard* requires that the following content is covered in training:

1. All hazardous materials in the workplace
2. Material Safety Data Sheets

3. Labeling
4. Health and physical hazards
5. Hazard assessment and effects of overexposure
6. Basic spill and fire response
7. Storage and handling
8. Protective equipment

RCRA's *Hazardous Waste Generator* regulations, Title 40 CFR Part 264, require this content to be covered:

1. All hazardous wastes in the workplace
2. Waste characteristics
3. Labeling and marking
4. Physical hazards
5. Spill and fire response
6. Communications/alarm systems
7. Equipment shutdown
8. Storage and handling
9. Protective equipment

On the surface, these two regulations have nothing in common. After all, they aren't even monitored by the same agency. But when you take a second look, it becomes evident that they are actually closely related. Focusing on the intersections can help both regulations make more sense. Where does the content overlap?

First, we need to clear up a problem of terminology. The Hazard Communication Standard clearly states that it covers only the hazardous materials in the workplace; it exempts hazardous waste. However, what is hazardous waste but *used* or *spent* hazardous materials? In the workplace, employees often don't know this. They have a confused view of hazardous waste because they don't understand what it is. They may know it is supposed to be labeled a certain way or stored in a special area, but beyond that, they don't know anything about it. They may, as a consequence, be unreasonably afraid of it, even more so than they are of hazardous materials. This confusion can lead to dangerous mistakes.

Sometimes the health and/or physical hazards of a waste are greater than those of the individual materials that created it; and sometimes they are less. An employee has to understand the nature of *both* in order to handle them safely.

Does it make much sense to train an equipment operator in the health and physical hazards of the virgin hazardous materials he or she handles and say nothing about the hazardous waste stored adjacent to them? No, it really doesn't, especially when you consider that the waste *is made up of* the hazardous materials. So, since hazardous waste and hazardous materials are inextricably related (most of the time you won't find one without the other), why not pull a little Right-to-Know into your hazardous waste training and push a little hazardous waste over into Right-to-Know?

Remember, the regulations are *performance-oriented,* they want *you to come up with an approach that works for you.* If you can make this mental adjustment regarding the relationship between Right-to-Know and hazardous waste, your training for both regulations will work a lot better and make more sense.

Instead of talking about OSHA labeling at one class and hazardous waste labeling at another class, which can be confusing to employees and come across as if you are contradicting yourself, why not hold one class on *container labeling* and cover both regulations?

Instead of holding a Right-to-Know class that covers how to clean up a spill of lacquer thinner and then holding a hazardous waste class on how to clean up a spill of waste paint–related material, thus giving the impression that they are two vastly different situations, why not hold one class on *spill response* that covers both?

Instead of covering container storage of flammables and combustibles with all employees in a Right-to-Know class and then discussing hazardous waste storage with a few select personnel at another class, wouldn't it make more sense to hold one class on *container storage* so that all employees will have consistent information on all the containers stored in the workplace?

Don't think you have to be limited to only one content area per training session; not at all. The amount of required content you cover at any given training class need only be limited by the amount of time available, the attention span of your students, and their ability to absorb and retain information on a variety of subjects when presented together. And these are factors only you can judge.

The Hazard Communication Standard and Hazardous Waste Generator regulations are not the only regulations with training content overlap. Both of these regulations intersect with the content of OSHA's Standard for *Employee Emergency Plans and Fire Prevention Plans.* As examined in Chapter 5, this Standard requires training in:

1. Hazard assessment
2. Spill and fire response
3. Communications/alarm systems
4. Equipment shutdown
5. Evacuation
6. Emergency/fire prevention

This Standard offers a combination of the emergency response and prevention elements of both the Hazard Communication Standard and Hazardous Waste Generator regulations.

At the most basic level, an employee cannot assess emergencies or prevent them from occurring unless he or she knows about the health and physical hazards of the materials and wastes in the workplace. Since you want all of this information to fit together in your students' minds eventually, it seems logical to combine it on the front end. In other words, from the time you begin talking about the hazards of materials and wastes in a Right-to-Know class, emphasize the *value of that information* in emergency prevention and response. Don't wait until a separate training session to mention it. The more you talk about the practical uses for the hazard information you give your students, the more likely they will be to listen to it and remember it. Giving them ways of using the information is part of selling it to them, making it *theirs.*

The content of the Hazard Communication Standard, with its emphasis on *hazard information* is so interrelated with Emergency Action/Fire Prevention, with its focus on *information application,* that once you give it a little thought, you will likely find it impossible to hold a class on one without talking about the other.

Once you move beyond the content areas that intersect with the Hazard Communication Standard, the rest of the Standard overlaps with the Hazardous Waste Generator requirements regarding Preparedness and Prevention and Contingency Planning. These content areas include spill and fire response, use of communications and alarm systems, equipment shutdown, and evacuation. The only difference between these topics, as presented in the regulations, is that, like the Hazard Communication Standard, the Standard for Emergency/Fire Prevention Plans doesn't mention hazardous waste. However, if *you* make the connection between materials and wastes and *you explain it to your students,* there is no reason why you cannot cover the required procedures for both in one training class (see Figure 8-1).

As discussed in Chapter 5, as much as you should combine training under these two regulations you should also combine your written programs. In fact, doing so will serve in making your training much, much simpler. Can you imagine the confusion that could be generated in having one evacuation plan (as part of your RCRA Contingency Plan) that took into consideration only the location of hazardous wastes in the workplace and another evacuation plan (in your OSHA Emergency Action Plan) that only addressed the location of hazardous materials? This really isn't that farfetched, especially if you consider that the OSHA Emergency Plan could be one manager's responsibility and the RCRA Contingency Plan could be another's. Combining procedures (whether written or not) is integral to the successful coordination of training.

Protective Equipment

Personal protective equipment is often the missing link that keeps an otherwise solid training program from having the measurable impact it should—and can—have. Perhaps this is because, while companies focus on compliance with the specific OSHA Standards for protective equipment (see Chapter 4), the relationship between protective equipment and the other training regulations is overlooked.

176   Safety and Environmental Training

OSHA 29 CFR Part 1910.1200	RCRA 40 CFR Part 264	OSHA 29 CFR Part 1910.38
Hazard Communication Standard "Right-to-Know"	Hazardous Waste Generator Requirements	Employee Emergency Plans and Fire Prevention Plans

Training must cover:
    All hazardous materials
    in the workplace
	Training must cover:
	    All hazardous wastes in the
	    workplace
		Training must cover:
		    All hazardous materials
		    and wastes in the workplace

MSDS	Technical data (waste characteristics)	
Labeling	Labeling and marking	
Physical hazards Health hazards Hazard assessment Effects of overexposure	Physical hazards Health hazards	Hazard/risk assessment
Basic spill reponse Basic fire response	Spill response Fire response Communications/alarm system Equipment shutdown	Spill response Fire response Communications/alarm system Equipment operations/shut down
Storage and handling procedures	Storage and handling procedures	Storage and handling procedures
Personal protective equipment	Personal protective equipment	
Reporting	Evacuation procedures Reporting	Evacuation procedures Emergency prevention

Figure 8-1.   Employee Training Requirement Cross-Reference Chart.

This relationship is critical because training regarding protective equipment is only valuable *in context.* And the regulations governing the specific hazards of the workplace *provide that context.* For instance, you can conduct a class on respiratory protection that addresses how to wear a respirator, how it protects the body from a biological perspective, and how to clean, store, repair, and replace it, without once mentioning the role of the respirator in your particular workplace. Presented in this manner, the information is *out of context* and, as such, will not be very memorable or valuable.

However, if respirators are incorporated into all training that deals with the hazardous materials and wastes at your company, then they take on a role that is meaningful because it has presence (really exists) and place (fills a need). For example:

1. The best time to bring up the use of respirators in the painting department is during a class on the health hazards of the chemicals in the painting department—not at some later class.
2. During a meeting on the proper procedures for pumping a solvent from a drum to a smaller container, go ahead and talk about the need for gloves and goggles during this activity.
3. During a training class on responding to spills and leaks, it makes little sense to discuss how to stop, contain, clean-up, and dispose of the spill, while leaving

out the equipment that employees need to wear to protect themselves in the process.
4. While holding a session with the employees who will be responsible for lagging behind in the event of an evacuation to maintain essential operations or shut down the facility, discuss what protective equipment they might need in order to do this safely, as well as where it should be stored.

- The *accessibility* of personal protective equipment along with the employees' *knowledge of the benefits* of the equipment and *comfort in wearing it* all are key factors in determining the success of your training program.

Attitudes

These could be called *intersections of policy,* although that sounds rather dull. Attitudinal intersections are the topics covered in training that carry the message of the company policy on safety and environmental compliance.

Such attitudes are the wings on which policy is put in flight, carried straight into the heart of the company, through the hearts of each employee.
Such attitudes connect each training class to some greater purpose, some definitive goal to which everyone can look and know how far they have come and how far they have yet to go.
Such attitudes close the circle, connecting personal safety to environmental health to community wellness to quality production to financial stability to personal safety.

Where do these attitudes or policies come from? They come from within your company. No one outside your company can tell you what they should be. They come as the answer to the question: *What are we doing, here, anyway?* Are you making a better product? Creating a healthier environment? Building personal pride? What's your mission?

Once you figure out what it is, use training to get the message across. Again, and again, and again. Until it's everyone's attitude. Until it's part of company culture. Until it's *policy.*

The Art of Documentation

Covering multiple requirements in one class won't help you with the nuts and bolts of your compliance *unless you document it.* Each regulation that requires training also requires some sort of written proof of training. Such "proof" can vary from an attendance record with a description of the topics covered in the class at the top of the page to a written test developed especially for the information covered, or a questionnaire administered orally by the trainer and signed by *each student.*

Although such documentation is required by law, don't forget that the *best proof* of training is proper behavior out in the workplace.

Documentation is an art because it takes a practiced hand to use it effectively, as well as to convince your students not to be afraid of it. The fact that documentation in any form *cannot be used against them* must be repeated to employees until they lose their fear of it.

Apart from its role in establishing compliance with the regulations, the purpose of documentation is to provide a tool for the trainer to use to judge his or her effectiveness *at the time of the class.* If you don't do the same thing all of the time, but change it so that it's unpredictable, you can use documentation to assess how closely your students are paying attention. A short "pop quiz" in the middle of a session can perk up flagging interest. Group activities, such as answering a list of questions based on a Material Safety Data Sheet or filling out a label can serve as documentation of the information covered in the class while they break up the monotony of verbal discussion.

Documentation can lend continuity to a training program. If you are meeting with employees on a weekly basis, a quick oral or written review of last week's topic before this week's meeting can provide a sense of the evolution of the program. The equivalent of saying: "Look, guys, we *are* going somewhere with these classes!" Documentation also can help link regulatory topics. For instance, at the end of a class on spill response, the trainer might ask the students to write down the health and physical hazards created by a spill of liquid sodium hydroxide. Then the trainer might follow-up with the question of what personal protective equipment they should wear to clean up such a spill. The trainer's notes on the class connecting spill response, traditionally a RCRA concern, to OSHA's Hazard Communication and Protective Equipment Standards, coupled with the employees' written responses to questions emphasizing this connection, will serve as meaningful documentation of coverage of at least three regulations.

And, finally documentation can be an effective method not only for determining how well your students understand what you're trying to teach them, but for providing *them* mile markers by which they can judge what they have learned. Documentation can be as much their tool as yours (see Figure 8-2).

REDEFINING THE SAFETY MEETING

Safety meetings. Production meetings. Department meetings. Job site meetings. Project meetings. You probably have some kind of on-going, routine meetings at your company. And chances are you're not using them to best advantage, because you're probably not using them as an opportunity to meet training requirements.

Case in Point: The Vice President of a company that specializes in emergency response, spill clean-up, and tank cleaning is concerned that they are not meeting

EXCELLENT MANUFACTURING COMPANY

Date: _____

The following personnel attended a ____-hour training class where the requirements of these regulations were covered:

         Citation   Topic

1.

2.

3.

4.

Instructor signature:_____

Attendees:

1.	11.
2.	12.
3.	13.
4.	14.
5.	15.
6.	16.
7.	17.
8.	18.
9.	19.
10.	20.

Figure 8-2. Training Documentation Form.

annual training requirements. In discussing the company training program with a consultant, however, it becomes evident that there are many times during normal operations when training requirements *are* being met. But nobody at the company has realized it.

When a clean-up crew arrives at a spill, the first thing that happens is they have a job site meeting. At the meeting, the supervisor issues personal protective equipment to each of the technicians. He inspects respirators and SCBA and conducts fit tests on each technician who will wear one. Before anyone begins work, the supervisor reviews the health and physical hazards present at the site and sets out a workplan that involves procedures for use of protective equipment and decontamination.

The consultant suggests that, since supervisors must fill out a job log at each site, it would be a simple matter to attach a piece of paper to the log where the supervisor could record the topics covered at the pre-work meetings. Such meetings, conducted and documented at job sites over the course of a year, could serve to meet the annual 8-hour refresher training required for the company's hazardous materials technicians by OSHA Title 29 CFR 1910.120. They could also meet requirements of Hazard Communication, Personal Protective Equipment, and Confined Space Entry training.

Adopting this system, the Vice President discovers that, for his employees, training that is on-going, and work-site-specific is much better received, and therefore more effective, than training conducted in an 8-hour-block.

If you have weekly or monthly safety or production meetings at your company, you have an organizational structure already in place that can help you accomplish two goals with respect to your training compliance:

1. You can make on-going learning about workplace hazards part of the company culture.
2. You can make training everyone's responsibility.

How can you use safety or production meetings to accomplish this? There are several ways:

1. *Develop a schedule of training issues to be covered at meetings*—Depending on the size and structure of your company, you might want to involve supervisors, department heads, or your entire staff in a "brainstorming session," the purpose of which is to develop a list of topics addressed in the training regulations impacting your company. Working from this list, you might cover one training topic (such as "Hazardous Materials/Hazardous Waste Labeling" or "Spill Control") at a safety or production meeting each week, every other week, or every month. One month is absolutely the longest you should wait between meetings where training-related concerns are discussed. The frequency with which you hold meetings will depend, in great part, on the nature of your operation and the level and variety of hazards your employees encounter.
2. *Build series of meetings around single themes*—Building several meetings around one theme aids in the reinforcement of a topic by addressing it from several angles or perspectives. It also helps to integrate procedures introduced

in the meetings more fully into the worklife. For instance, if the training theme for one month of safety meetings (one held each week) is "Spill Control," it can be addressed from a slightly different angle, and *meet a different regulatory requirement* each week.

Week 1: Cleaning up hazardous materials spills (complies with Right-to-Know).
Week 2: Cleaning up hazardous waste spills (complies with RCRA Hazardous Waste Generator Requirements).
Week 3: The right personal protective equipment for the job (complies with OSHA Personal Protective Equipment and Respiratory Protection Standards).
Week 4: Minimizing waste by avoiding leaks and spills (complies with Waste Minimization regulations).

It is easy to see how, spending just 30 minutes a week for four weeks, you could provide two hours of spill control training for your employees while addressing four regulatory requirements. And just think, you probably already have the organizational structure within which to do this.

3. *Let different employees run the meetings*—Regardless of the size of your company or the type of operations you conduct, the quality of your meetings will be enhanced by letting different people run them. Every person in your organization has a different perspective on every safety concern. Even if more than one person does the exact same job, each brings a different background of experience to it. Such experience is invaluable to your training program—but only if your employees have the opportunity to share it. Remember the oral book reports and speeches you had to deliver in high school? Something about translating information or knowledge into spoken language makes it last longer, doesn't it? When your employees talk about safety in front of their coworkers, their knowledge is not only brought to mind, *it is validated.*

9

A Little Respect

If you disrespect
Everybody that you run into
How in the world do you think
Anybody's s'posed to respect you?

> "Respect Yourself"
> The Staple Singers

Whether you stand at the breakroom door preparing to walk into your first training class or your four hundredth, sooner or later there will come a time when you feel desperately alone. And this sense of aloneness is nearly impossible to escape. Because even if safety and environmental policies and procedures have been integrated into those of total company management, and even if every other manager in the company gives you a great deal of support, when you hold a training class, you are emotionally, intellectually, and physically alone.

It's at moments like these when you're likely to wonder how in the world you got yourself into this. It's at moments like these when you realize, with a force like gravity, that it is one thing to stand in front of a class of people and cover the requirements of a regulation, but it is quite another matter to teach them in such a persuasive manner that they are inspired to change their behavior. Sometimes the latter task seems much too great, especially for a mortal with other things on his or her mind. But training is, nonetheless, your job.

With only dry, technical material as guidance and your own wits as tools, you must inspire, motivate, and educate. You must convince your fellow employees to change their attitudes, broaden their perspectives, and look at their jobs in a whole new light. Ultimately, you must alter their behavior. You must lead a one-person blitzkrieg on habit, tradition, and "the way things have always been done," and somehow manage to do this without offending or alienating anybody.

Ultimately, you must seduce your students to a new way of thinking by making it so attractive that they find themselves, quite literally, unable to resist. You must convince them to do what's good for them in the long term, even if it makes them uncomfortable right now. You must take what they will perceive as "bad news" and make it not only appealing, but irresistible. You must transform boredom into excitement.

And you must do all of this patiently, persistently, a little bit at a time—always watching, listening, and observing behavior; because the degree to which it changes is the only yardstick by which you can measure how well you are doing your job.

So how do you do it? How do you summon the energy or courage, the gall or chutzpah to keep walking into that room and opening your mouth?

Is there any one recipe for success in training, for accomplishing the goals you set out to accomplish? Is there a step-by-step procedure that works across the board? Is there an "insert flap A into slot B" kind of formula? A certain type of tool: videos, posters, booklets, slides? Special gimmicks, tricks, showmanship, oratorical skills? Something simple you can go out and buy?

No. There isn't.

What will make your classes succeed is very much akin to the passion that must drive the entire program—and that is a deep concern for fellow human beings. A sense of responsibility for their well-being. A caring consideration for their point of view.

A little respect.

This respect cannot be bought and it certainly can't be learned by reading a procedural "how-to" manual. It must be found. You must find it *inside yourself.*

Find it? you ask. Is this some kind of psychic treasure hunt? Actually, yes, it is something quite akin to that.

Assuming you have moved beyond the "I have to, it's my job" frame of mind, you have your own unique reasons for being a trainer. You have your own reasons for thinking that the information you're charged with teaching needs to be taught. You have reasons for caring about your fellow man that are based on your personal experiences, both good and bad. You have special knowledge to share, insights unique to you, that only you know about.

The irreplaceable nature of *your perspective* is what will make your training not only successful but sustainable. It is what will make your fellow employees want to listen to you. This honesty that says that you respect them enough to bring them into your confidence, to give them a little of yourself.

You see, though you may be surrounded by expensive teaching aids and fancy gadgets, when you stand in front of a room of people without having first searched your heart and asked yourself, why am I here, then you stand exposed, vulnerable, and pitifully empty-handed. There is no meat to your words, no substance to your warnings, and no compassion in your demands that your students change.

And they can tell. You don't respect them enough to be real.

The irony is that, by exposing yourself, by revealing your motivations for training, by flat out telling them you care, by engaging in what the world would describe as reckless, dangerous self-revelation, you erect a strange kind of forcefield around yourself. And, eventually, if not right away, your care for your students is returned to you.

But it takes time to build this kind of trust. This is not a miracle remedy. It won't happen overnight. When you take this approach it is almost guaranteed that there will be those individuals in your classes who will be threatened by it. The usual condescending, authoritarian approach is what they have come to expect, even depend upon. And when you walk in with your honesty and your "we're all in this together" team spirit, looking them in the face and challenging them to change, some of them will respond with an aggression that will attempt to discredit and undermine your very presence in the room. You can just about set your watch by it.

And this is exactly when it is so critical that, while you hold your position, you don't get angry. Every moment you respond to their aggression sanely, straightforwardly and, whenever possible, with humor, that forcefield gets just a little taller and a little stronger. Their respect for you grows. But if you meet aggression with aggression, then it's them against you. And you don't need a bookie to tell you that in that situation you're the odds-on favorite to lose.

If you fight fire with fire, soon the whole world's ablaze. This doesn't mean, however, that you have to become fuel. It means you must work to outsmart the fire, to find the best vantage point from which to douse it with cool water and *put it out*.

Just think of it from the employee's point of view: After awhile it simply ceases to be fun to try and pin the tail on a down-to-earth person who's obviously trying to teach them something. But the sport of pinning the tail on a pompous ass lording it over them about how much he knows? That's one sport that never loses its allure.

So, when all is said and done, it is respect that will save you, and sustain you, and, perhaps, keep you from becoming discouraged. It is respect that will provide you the only tools you really need to accomplish everything you want to accomplish through training. And those tools are *humility* and *voice*. Out of respect, you humble yourself to reveal your reason for being there. And through that self-revelation, you find a voice that brings your training to life, that makes what you say worth listening to, because it's true.

It doesn't come out of a Code of Federal Regulations; it comes out of the real world.

So, that said, here are a few tried and true methods for using respect, humility, and voice to make your training classes come alive.

GIVE CREDIT WHERE CREDIT IS DUE

What you want
Baby I got it
What you need
You know I got it
All I'm asking
Is for a little respect

           "Respect"
         Aretha Franklin

No one likes to be talked down to. No one responds well when they feel as though someone who has never done their job is telling them how to do it.

It doesn't matter whether you are the new kid on the block who was recently hired to handle safety and environmental compliance, or whether you have been at the company for years and were drafted into the position, there is going to be a certain amount of prejudice against you, and a certain amount of skepticism about everything you say. There are going to be a lot of your fellow employees looking at you with their hands on their hips and their heads cocked to one side thinking that you're taking yourself far too seriously.

You might as well get used to this and not let it bother you because, at least at the beginning, you can't get around it. If you're the new kid just out of college, they've got your number because you're inexperienced. There is a whole lot more you *don't* know than you do know. The same thing is true if you have experience, but it was gained at another company. You're new to them. Even if you've worked in a non-production-oriented area of the company, such as the personnel department or the lab for 25 years, they're going to look at you as the new guy. They're not going to want to listen to you talk about safety, because, after all, what do you know about what they do? And, perhaps most difficult of all, if you've been involved in production as a foreman, supervisor, or manager, and now "all of a sudden" you're holding classes trying to get your people to follow safety procedures, they're going to remember every bad habit, every shortcut, and every prior safety violation you've made or watched them make. They're going to want to know why, if it was okay then, it isn't okay now. And they're going to hold you to it.

The best way to defuse this attitude is to disarm your students by humbling yourself before them. There is no trick to it; *just be honest.*

You're the new hire—whether straight out of school or from another company. You're standing in front of a room full of machinists and welders, but it could be anybody: printing press operators, mechanical platers, painters. The point is, you've never done their job. So, tell them what you don't know, but emphasize your interest in learning. In addition to surprising and flattering them, this dialogue will accomplish several things. It will set their pride at ease by establishing that what they do is important, requiring special knowledge, skills, and experience, all of which you appreciate. It will set the tone of your classes as that of an easy-going conversation and information exchange rather than a didactic lecture. And it will teach you a lot, not only about the work your students do, but also about their attitudes toward their work, their habits, and even their philosophies.

One Approach: "I want to start out by making a confession: I know a lot about safety, but I don't know much about machining and welding. That's your area of expertise. If you are willing to teach me about what you do, then I think what I say about safety will be a lot more valuable."

People love to talk about what they do. Ask questions. Even if you think you already know, *ask*. Hear it in their words. Go out into the shop and have them show you their equipment, have them demonstrate how they really do things. By your interest and curiosity, you are giving them credit for knowing something you don't. And they will come to like and admire you for it.

The premise is basically the same if you've been transferred or drafted into safety from another area of the company. If you really don't know much about production, you might want to follow the exact same approach as above. But, if you've been around a long time and you feel you have a pretty solid understanding of production operations, you might take a slightly different tack. You might enlist your students as critics of your knowledge of what they do. This kind of self-effacement will be startlingly refreshing to your students, especially coming as it does from a "superior." If you work at a company where you've known your students for years, this can be a lot of fun. Use your knowledge of personalities, friendships, and gripes, as material to work from.

One Approach: "I've known most of you a long time, and I know you think I don't do anything all day put push paper from one side of my desk to the other. You guys don't think I know the first thing about what goes on out in the plant. Am I right? Well, in the safety meetings and classes we're going to have over the next few months, it's my goal to surprise you. But just to make sure I don't get to far off base, I'm going to ask you to do something I know you've been wanting to do for years. I'm going to ask you to be my critics. And, what's more, I'm going to shut my mouth and listen to whatever you say."

As a Production Manager, you are in a sticky situation where safety training is concerned because your students know you so well. They know your bad habits. They know your track record on safety. There is no pulling the wool over their eyes. So one word of advice: don't try.

Being a Production Manager is a disadvantage where safety and environmental responsibilities are concerned only if you adopt a holier than thou attitude and try to pretend that your students don't know you, or if you act as though you are better than you are, which implies you think you're better than them. No one likes insincerity, especially in a leader who they are supposed to trust. The minute you do this, you've lost all credibility. And you'll probably end up losing friends and admirers.

But, if you avoid this trap, managing production affords you a unique and valuable perspective on safety—as long as you're consistent. And consistency is difficult. You are human. You have production deadlines to meet. So go ahead and admit to your students that you're not perfect. Give credit to those who have good, consistent safety habits and enlist their help in monitoring and reminding you, keeping you on the straight and narrow. Give credit to those of your students who team well out on the production floor, and tell them how important that teamwork is in safety.

You know your students better than anyone else in the company. Recognize their strengths. Build your classes around them.

One Approach: "Doing your job safely is the only way to do it properly. I see some of you are surprised to hear me saying that. You might have expected me to say that doing your job quickly and efficiently or with the fewest number of errors or flaws in the final product is the only way to do it right. And that's what I would have said in the past. And all of those things are important. But none of them amount to a hill of beans if one of you gets hurt on the job. I have changed my attitude, people, and I am asking you to help me stick by it. If one of us in this room allows anyone else in this room to put production concerns ahead of safety, then we aren't doing our job properly. Because unsafe production is not quality."

REMIND THEM WHAT THEY ALREADY KNOW

It is not unusual for employees to be a little nervous when they walk into a training situation. This nervousness can manifest itself as defensiveness and aggression, as clowning silliness, or as an aloof refusal to participate. Sounds a little like the behavior patterns of kids on the first day of school, doesn't it? Each individual masks their insecurities with an adopted persona worn like a mask. And who can blame them—the school kids or the adults in your class? It takes an extremely strong personality to walk into an unknown situation and "be yourself."

So how do you deal with this? First, you need to get a handle on what your employees might be afraid of.

The Literacy Factor

If an employee has poor literacy skills, then anticipated reading or writing during class will be feared.

If many of the employee's coworkers also have poor reading skills, then he or she probably isn't afraid of discovery, but might be afraid of losing his or her job or otherwise being punished for not performing well in the class. The fear of punishment can be so great that it totally distracts the employee from what you are saying, thus insuring that, in fact, nothing is learned in the class.

When an employee is trying to conceal illiteracy, the situation is more difficult because, as the trainer, you have no way of knowing what the problem is. If the employee is afraid of discovery, he or she will likely do anything to avoid the class. When the employee comes to a class, he or she may try to overcompensate for a lack of reading skills by talking and asking a great deal of questions, or he or she may be highly belligerent and refuse to talk at all. The employee may cheat on written activities or tests or, in defiance, turn in blank papers.

The Age Factor

Older employees may be afraid of the classroom situation simply because they haven't been there in so long. Here they are, after 30 or 40 years at a job and they have to sit in a class and learn new ways of doing what they've been doing for more than half their life. Bah humbug. They're afraid of change. They're also afraid of being revealed as old-fashioned, outmoded, or not knowing what it takes to keep up with the newfangled style of doing things. These employees are likely to be really, really *quiet.* Although sometimes they may burst out with a tirade that goes something like: "I remember how things used to be when all a man had to do was a good job. Whether or not he was safe was his problem. After all, any one of us can get hit by a truck on the way home."

The Higher Education Factor

Employees with college degrees don't seem afraid of training classes, do they? No, they just seem above them, too busy for them, or unaffected by them, which is, perhaps, just another mask for fear. Do the managers, chemists, or engineers on your staff participate in the training program? If they don't, perhaps it is because they really *do* know everything that the training covers (in which case it certainly would be no skin off their nose to sit in on some classes and find out). Or perhaps it is because they are afraid they *don't* know everything and they certainly don't want this weakness revealed to other employees. Or perhaps it is because they are afraid that they will not have a chance to prove how much they know and how superior in intellect they are to everyone else.

Putting Fear to Rest

Strangely, one approach works to put all three of these fears to rest. Assure your students that you are not going to discuss anything that they don't already know or haven't already experienced. Tell them that the purpose of the class is to *share* work experiences and knowledge. By examining accidents, injuries, fires, and explosions and figuring out *why* they occurred, everyone can learn to work more safely. For example, in a Hazard Communication class, first-person tales of childhood pranks with aerosol cans or gasoline, or adult mishaps with charcoal lighter fluid, WD-40, hairspray, or bleach and ammonia serve as the personal, almost visual illustration upon which the class can be based.

A dry explanation of volatility, vapor density, flammability, and flashback is greatly enhanced in the context of Mary's story of her brother using gasoline to clean paint brushes in the bathroom sink...while a space heater was turned on. Or Marvin's story about catching himself on fire after washing his arms and hands in lacquer thinner...and then lighting a cigarette. The importance of proper storage and handling of compressed gases is vividly illustrated when Leroy, a seasoned welder, described the time he saw an acetylene canister get knocked over by a forklift and shoot off like a rocket, crashing through a wall.

And the critical importance of developing and following safe operating procedures is brought to life when Cheryl describes how, when working as a bartender at a hotel, she poured an acidic greasecutter down the bar drain before leaving for the night. The next morning, a janitor added drain opener and hot water. After that, a housekeeper, noticing a strange odor coming from the bar, added bleach to cut the smell. By the time Cheryl returned to work, the sulfuric acid vapors emanating from the drain were so strong that she passed out and had to be taken to the emergency room. The sink and pipes had to be replaced.

In a class where literacy might be a problem, emphasize that the purpose of the class is not to discover what they *don't* know, but build on what they *do* know. Tell them that no one is being graded on their performance in the class. Tell them that there are no wrong answers, that everyone's insights, observations, and experiences are important. In fact, the only "wrong" thing they can do is not participate.

Where older employees are present, emphasize the *value of their experience.* Encourage them to talk about what they have learned since they started in the business. How have operations changed? Are they safer? Encourage them to describe accidents, injuries, or other emergencies relevant to the topic of the class that they have observed. Why do they think the accidents happened? How do they think they could have been avoided?

When managers or staff members with technical degrees attend a class, put their egos at ease by giving them the opportunity to *share their expertise.* Regardless of the topic of the meeting, whether it is chemical or environmental hazards, energy hazards, maintenance procedures, or protective equipment, it is doubtful that they know "everything" about it because they probably have looked at the subject from only one perspective: *theirs.* If you use them in your meeting as your technical resource—your resident "experts"—they will be happy because they get a chance to show off.

And as you relate their technical knowledge to the work experiences recognizable to the other employees in the class, everyone will learn something. The experts will gain a new perspective on their knowledge through its practical application. The other employees will come to understand that all the technical "mumbo jumbo"—the facts, figures, and statistics of safety—have a direct relationship to their daily on-the-job experiences.

Focus on Why

When you breathe solvent vapors, they make you feel "high." Everyone knows this. If they have ever worked with paint, thinner, wash-up solvent, degreaser, or gasoline, they've felt it. But what they might not know is *why*. Solvent vapors make you feel high because, when you breathe them, they enter your blood stream and travel to your brain, affecting your entire nervous system.

If you leave a can of acetone exposed to the air, the acetone can be ignited by a heater across the room. Many people know this because they have seen it happen. But *why* does it happen? What they may not understand is that it is the acetone *vapors* that catch fire. These vapors are heavier than air, so as the acetone evaporates, the vapors gather on the floor and move about the room, seeking an ignition source. When the vapors meet the heater and are ignited, the fire travels back along the trail of acetone and ignites the can. The mandate "Keep lids on containers of solvents" takes on new life and meaning with an explanation of *why*.

Why does one corrosive liquid make a slick, soapy feeling on the skin while another burns right away? The man or woman on the street probably doesn't know that some corrosives are acids while others are bases; do your employees know the difference? Bases, like caustic soda or lye, set up a barrier on contact with the skin, dulling pain and allowing for a longer exposure period. However, since the base reacts with the fatty acids in the skin, the ultimate result can be more serious burns than those caused by contact with certain acids. All corrosives do not behave the same because they are chemically different.

Tap your students' curiosity. Explain that once they understand why a chemical behaves the way it does, they will be better prepared to handle that chemical safely. Once they understand why a fire has occurred, they will be better prepared to prevent fire in the future. Once they understand why any given hazard exists, they will be better prepared to avoid or minimize it. This knowledge, rooted in their experience, gives them an autonomy over their lives that ignorance has prevented.

Your students will be empowered by the realization that fires, explosions, chemical reactions, environmental impacts, health effects, and accidents are not random events over which they have no control, but predictable and preventable occurrences. And the first step in predicting and preventing them is to examine their own experience.

UNCOMMON COMMON SENSE

Case in Point: Moments before a training consultant begins a Right-to-Know class at a machine shop in rural Mississippi, the Vice President of Operations takes him aside and whispers: "I've got to warn you, they can be mean."

The trainer raises his eyebrows.

"Lemme put it to you this way: If they aren't mean, most likely they're asleep."

The trainer takes a deep breath. This is just the kind of wisdom he likes to hear as he prepares to spend two hours standing in front of 40 rather grizzled machinists to whom he is supposed to impart some knowledge of basic chemistry. Still, the show must go on.

After the V.P. introduces him, he walks back and forth across the front of the room, trying to establish eye contact. "How many of ya'll don't want to be here?"

There are a few surprised grunts as about half the group raise their hands.

"How many of ya'll are bored already and plan to stay that way?"

Now everyone raises their hand and a few men actually crack a smile.

"Well, I've got news for you." He points at several people. "Today I'm going to mess up your plans. You're gonna get so interested in what I say that you're gonna forget all about being bored."

He keeps walking back and forth across the front of the room and down aisles between tables.

"Somebody tell me why we're having this class. You, Steve, tell me."

"I dunno. 'Cause it's the law. 'Cause we have to."

"Okay, that's one reason. But there's a better one. What's the subject of this class?"

"Chemical hazards," someone else says.

"Okay, good. Who figures they know everything there is to know about the chemicals ya'll work with?"

Everyone raises their hand.

"Okay, great, that means I'm sure as we start talking about these chemicals, ya'll can teach me a few things. But let me ask you one more question, listen up: How many of ya'll have *never* made a mistake, *never* had an accident, or *never* made a bad judgement at any time in your life?"

No one moves.

"See, today I'm not here to tell you how to do your job. I'm here to talk about preventing accidents and protecting your health. And the best way to do that is to use common sense."

He's still walking back and forth. No one's asleep yet.

"How many of ya'll think you've got pretty good common sense?"

Most of the men raise their hands. A few make quiet jokes about each other and laugh.

"Well, guys, today you're gonna have a chance to prove it. I'm gonna put your common sense to the test."

There is at least one thing, besides birth and death, that all humans share—a desire to do the opposite of whatever we are told. If a label warns us not to mix a product with anything but water, we share an urge to play Mr. Wizard and mix several products together...just to see what will happen. If a label informs us not to smoke while using a product, few pause before lighting up again. If a label instructs users to wear goggles or gloves, we get in a hurry or simply forget.

We know that lids *should* be put back on flammable liquids right after we use them. We know that machine guards *should* be replaced after they're removed. We know that protective caps *should* stay on compressed gases. We know we *should* clean and store our respirator properly. We know that those drums of flammable liquids really *should* be grounded, and that spills that occur in the storage room *should* be cleaned up right away. We know that it isn't wise to use an air hose to blow dust off our clothes and out of our hair. We know that we should never clean moving parts while they're in motion. We know that chemicals should never be left in unlabeled containers.

Sure, we've heard all this stuff. But who has time for such details? We know what we're doing, right? We can afford to take shortcuts and ignore warnings because, after all, we're experienced, we're in control.

Common sense is none too common in the American workplace and household. Don't kid yourself that your company and employees are vastly different. They're not. Even if they take the time to read a label or a hazard warning, for some reason they won't necessarily believe it. Whatever it tells them *not* to do, they will feel almost obligated to do anyway. At least once. *Just to see what happens.*

Talking about the absurd safety risks that you and they have taken "because you weren't really thinking, you weren't using your common sense" is an effective method for breaking the ice, loosening up the class. Is there a barbecuer in America who hasn't, in a fit of impatience, squirted yet more lighter fluid onto already hot coals? Is there a male over age 16 who hasn't thrown an aerosol can into a bonfire? Are there a few folks who have poured gasoline into the carburetor while someone else cranked the engine? Has anyone ignored the "Don't Mix With Anything But Water" warnings on most cleaning solutions? Has anyone poured used paint thinner down the kitchen sink or poured used motor oil on weeds?

Everyone has a story. Some are funny. Some are harrowing. Each one has a moral: No one is so experienced, so old, or so educated that they don't need to be reminded, first, that they have common sense, and second, that they know better than to ignore what it tells them.

It is not the chemical or equipment manufacturer's fault or the boss's fault if an employee chooses to use chemicals, machinery, or equipment improperly. It is each individual's responsibility to respect the power inherent in the materials and equipment he or she works with and to use them in the manner for which they were intended.

THE POWER OF LISTENING

No one cares to speak to an unwilling listener. An arrow never lodges in a stone: often it recoils upon the sender of it.

St. Jerome, from Letter 52

There is crucial power in asking and listening. If it doesn't come naturally to you, don't panic. It can be learned. If you're willing to wet down your ego and swallow your pride, becoming a listener will be well worth the effort.

The trainer who listens is the trainer who, in time, will be privileged to hear his or her students' true voice. Hearing that voice is critical to the training process. If your students do not speak, and speak honestly, about their experiences, problems, and ideas, you will never know if you are reaching them. And unless they *know* that you are listening to them and taking them seriously, they will not tell you the truth. They will not risk letting you into their world if they have any inkling that it might backfire on them. You can bank on this fact as sure as the sun rises in the morning.

But no matter how earnest and concerned you are, there are some relationships that simply won't be rushed. Becoming a trusted listener is one of them. Be patient and persistent. Most importantly, just *be there*.

In their quest to know whether or not you are trustworthy, your students may even test your listening prowess and your trust in what they say, by leading you on a few wild goose chases. It has happened that, rather than tell their trainer what is really concerning them and thus expose themselves, employees have made up stories about problems with process equipment or respirators, just to see what would happen. Just to see how the trainer would handle it. Would he or she listen to them and act on their comments? Would the trainer say he or she was going to do something and then forget to follow through? Would the trainer try to get the employees in trouble? What would he or she do?

If the trainer is a good listener and consistently does whatever he or she says he or she is going to do, it may take only one or two trial runs and false starts before the students start speaking truthfully.

If the trainer who is *trying* to listen runs into this much reticence, where does that leave the trainer who only has ears for him or herself? Siberia is probably more hospitable, a room full of granite perhaps more expressive.

Bore: A person who talks when you wish him to listen.

                        Ambrose Bierce, from *The Devil's Dictionary*

We've all been there. In a seminar where the speaker's unconcern for the audience is only surpassed by his or her condescension, there is nothing anyone could say that he or she would stoop to hear. In a class where the teacher is so overbearingly superior we are terrified to ask a question; if we dare open our mouths, there is that slight possibility that he or she might turn us to stone. Or in the company meeting where the person talking is so all-fired-sure he or she is right that he or she never asks for anybody's opinion or let's them get a word in edgewise when they try.

194 Safety and Environmental Training

Case in Point: "Today I'm going to tell you about Health Hazards. OSHA has defined seven health hazard categories in the Hazard Communication Standard. These categories are: Toxic, Highly Toxic, Irritant, Sensitizer, Carcinogen, Corrosive, and Target Organ Effect.

Starting with the term toxin. OSHA defines a toxin as a chemical for which there is an LC-50 or lethal concentration dosage of 50 percent. This means that there is a concentration of this material that, on the basis of laboratory tests, is expected to kill 50 percent of a group of test animals when the material is administered in a single exposure, usually for one to four hours. Toxins are also called poisons. Some examples of toxins include poisonous gases, such as chlorine, some organic solvents, and many pesticides. Material Safety Data Sheets include information regarding the toxicity of a chemical under the TLV or PEL column, usually found in the ingredients section. A TLV is a Threshold Limit Value. A PEL is a Permissible Exposure Limit. A TLV is an exposure level under which most people can work consistently, day after day, with no harmful effects. TLVs are established by the American Society of Government and Industrial Hygienists. PELs are basically the same thing, except that they are established by OSHA. TLVs are expressed in parts per million, which is a unit for measuring the concentration of a gas or vapor in a million parts of air. PELs are usually expressed as a TWA, time weighted average or maximum concentration exposure limit, in the case of skin exposure. It is important to review the TLVs or PELs of materials you work with in order to assure that you are within recommended exposure limits. Any questions so far? Good. OSHA defines an irritant as..."

Sound familiar? Makes you want to groan just reading it, doesn't it? Nothing in this litany is geared toward the employees listening to it. It is painfully obvious that nothing is being said with them in mind. Evidently, the trainer does not intend for anything he says to have practical value to his students. He babbles a string of technical garbledegook with no concern for whether or not it can possibly mean anything to them. His only concession to "discussion" is a quick "Any questions? Good." at the natural break in the lecture. The presentation is so intimidating and the content so mind-numbing that, quite expectedly, no one makes a peep. And so the drone continues.

We've all been bored. As trainers, what we must strive never to be, is *boring*. Why are people boring? Because they don't listen. But even more than that, they don't bother to *ask*. By their total self-absorption, boring people let you know, one way or another, that your presence is utterly incidental to the fact of their presentation. Your presence is superfluous, a detail. You don't matter. Why should they ask you what you think, much less listen to your answer?

If you're a trainer, being labeled boring is the kiss of death. How do you avoid it? First you can try a few self-monitoring tricks during your training classes. How often do you ask a question? How much of the class time is spent in discussion?

Does everyone have a chance to talk? Do you ask your students if *they* have questions? Do you allow their concerns to determine the direction of the class? You might even try tape recording some of your classes and evaluating them. Be critical. Ask yourself: Is this something *I* want to listen to? You might recruit a friend who knows nothing about the subject to listen to a tape and tell you what he or she thinks. If your friend finds it interesting, that's a pretty good indication you're on the right track.

Becoming a good listener and avoiding being boring demands a level of self-reflection and observation that isn't easy. You have to take your ego out of it and see yourself as others see you. Your job is not to prove how much you know. Neither is it to win a popularity contest. Your job is to improve your company and the lives of its employees, through training and listening.

> *He listens well who takes notes.*
>
> Dante Aligheri, from *The Divine Comedy*

Asking and listening, observing and remembering, are the only means at your disposal for finding out what is really going on out on the production floor—where the problems really lie, which procedures are followed and which are not, and which procedures are on target and which miss the mark. This is the only way you can stay "up" on things. You don't want to be a fuddy-duddy, a nerd, yesterday's news. You don't want to be someone that workers say one thing in front of and something else as soon as you leave the room. So spend more time listening than talking. Ask. Observe. Write down what you hear and see.

And follow-through. If someone asks you a question, make sure and find the answer. If someone expresses interest in trying out a new kind of protective equipment, make an effort to get it for them. This is not to suggest that you give your students carte blanche to take advantage of your generous nature. It is to recommend, however, that you give your students the benefit of the doubt. Take their requests seriously. Prove it by following through, one request at a time.

CHANGE, CHANGE, CHANGE

> *The only sense that is common*
> *in the long run, is the sense of change.*
> *And we all instinctively avoid it.*
>
> E. B. White

You are introducing a new procedure through Hazard Communication training, such as wearing a full-facepiece respirator during an acid spill clean-up. Or, through

Lockout/Tagout training, you are introducing the procedure of locking out power at the main cut-off before replacing a roller on a press.

This news is going to go over like a lead balloon. No one in the class wants to follow these new procedures because they have always done things "the old way" and see no reason to change. You must provide a *reason to change that they can understand,* that is meaningful to them.

You could begin by asking an employee in the room who wears glasses when he or she started wearing glasses. Ask if he or she liked wearing the glasses at first. Were they comfortable? Did they seem to get in the way? Then ask why, if the glasses were so bothersome, the employee continued to wear them? After looking at you like you are crazy, he or she will probably say, "So I can see." Seeing was important enough to prompt the employee to change his or her lifestyle and wear glasses, even though the glasses caused discomfort at first.

The "old way" in this case was not necessarily the best way.

A similar experience can prompt people to quit smoking or drinking, diet, reduce sugar or cholesterol intake, or wear seat belts. In each case change usually occurs after a powerful, emotional realization that there simply *is no choice.* The doctor says if you don't stop smoking you will have cancer in five years. A loved one who never wore a seat belt is killed in a car wreck.

There is a precedent for changing behavior in nearly everyone's life. Upon these precedents you can build a case for changing work behavior.

Your students will stop saying "I can't change the way I do my job" when they realize two things.

1. They are capable of change, even if it causes discomfort and "takes some getting used to."
2. The "new way" of doing things has real, long-term benefits.

Change is hard for everyone. Pretending that it isn't doesn't do anyone any good. If what you say respects the difficulty in change while emphasizing the need for it, then it has meaning to individual lives. Your students not only listen, they are eager to hear more.

Case in Point: "With the proper motivation, you can get used to anything. Am I right?"

The trainer is standing at the end of a table in a tiny lunch room at an auto body shop. Crowded around the table are 15 painters and body men, all with at least 10 years experience. Several of the men are absent-mindedly fingering a selection of air purifying respirators spread out in front of them. Though it is obvious that they are listening carefully, their expressions betray a good-natured skepticism.

The trainer can almost hear them thinking: "What he says sounds good, but it's just not that simple. Besides, he doesn't have to wear one."

Suddenly the Shop Manager, who is standing in the corner, leaning against a snack machine, begins speaking: "There is not one of you in here who doesn't know someone who has died of lung disease that can be directly traced back to breathing this stuff. These are the facts people. And you know them. Hell, several of you worked with Al Crawford and he died at what? 45? And then, of course, there's my own father. Wayne, here worked with him. But we keep on thinking it can't happen to us."

He picks up a respirator. "I don't know. I'm asking: What more motivation do you need to wear one of these things?"

The room is silent.

Finally, one of the men looks at the trainer and says: "Every year after one of your classes Terry here is all gung ho about his respirator. We laugh at him, you know. We kid him and say things like: So, you're gonna turn over a new leaf this time, huh, Terry? And he says: Yessir I'm gonna do it this time. I'm gonna wear it from now on."

The man nudges Terry in the shoulder. "How long'd you wear it last time he came?"

"Bout six weeks." A few men giggle and shake their heads. Terry's staring at his fingernails. "But that was about three weeks better than I did the time before."

MAKE IT MATTER

Case in Point: "I want to make a difference at this shop," says Claude, Production Manager at a medium-sized commercial printing company. "I want to have the best OSHA program in the city. And if I push hard enough, I can get the money to do it."

He shakes his head, confusion evident in his eyes. "But I go out in that shop and ask for feedback. I talk to the supervisors. It's obvious no one cares."

Claude's intentions are above reproach; he wants to initiate change and improvement through compliance. But his methods are misfiring.

Many high school graduates who made it through an English class of diagramming sentences or an explanation of symbolism in Hawthorne's *The Scarlet Letter* wondered, "What does this have to do with real life?" But most remember (with fondness) the teacher who brought history alive by relating it to some aspect of life experience.

Whether standing before a room of tenth graders or veteran truck mechanics, you cannot teach what does not matter. If it is not relevant, the information will not sink in—regardless of its inherent value or the sincerity with which it is presented.

For this reason you must build your training classes around the hazardous materials, machines, equipment, and safe operating procedures at your site. When standing before the class, take into consideration education levels and even student personalities. If your goal truly is to teach them, then everything must relate to *their individual lives.*

Because of the sex, age, education, cultural background, and work activities of its members, every work force is unique, requiring different information presented in various formats. At a time when training videos abound, be careful not to fall into the standardization trap.

If you run a printing company in Memphis, Tennessee, the makeup of your work force is different from that of a machine shop in Elmhurst, Illinois. Even the accents and idioms of video narrators will impact each site differently. Likewise, if you operate a vehicle maintenance shop, the mechanics probably will not learn from Hazard Communication videos based on materials at a manufacturing company. A machinist or welder tunes out a video about hazard labeling in a laboratory when he or she is concerned with cutting oil and welding fumes.

People simply do not care about information that has no bearing on their lives. Why should they? Standardized training rarely, if ever, is as effective as training developed specifically for your work force.

Case in Point: "Health Hazard. Tell me, anybody, what makes something a hazard to your health?"

"It's bad for ya."

"Okay, good. Something is a hazard to your health if it's bad for you. But how do you know it's bad for you? How can you tell?"

The trainer walks around in the middle of a warehouse-like plant talking to a mixed group of cutters, welders, and painters. They're sitting on boxes, pallets, and a few folding chairs. Although all equipment has been turned off for the meeting, he has to talk at the top of his voice to be heard in the cavernous room.

"You feel bad."

"You get sick, something goes wrong with you."

"Hell, you might even die."

"Right, you're all right. When we're around certain chemicals that can cause some change in our bodies, that can make us feel bad or get sick, we say those chemicals are poisons. Another word for poison is toxin. Who has heard the word toxin before?"

A few people raise their hands. "Who has ever been exposed to a toxin or experienced a toxic reaction?"

No one moves.

"Okay, let me put it to you this way." He stops, scans their faces as he asks slowly, "Is there anyone here today who has ever gotten drunk?"

The men smile, giggle, and nudge each other. "You know what I mean. Have you ever gotten a buzz, or just a little tight, or high. Even if it was just once. Do you remember what it felt like?"

"Once! Hey Jerry, ask him if he means once a day!"

"Otis, tell him why you got them shades on, man!"

"Okay, okay! Listen up! I need to ask somebody a serious question." The group quiets down. "Jerry. Yes, you. Describe for us what it feels like to get drunk. Go on, you can do it. Speak to us from your experience."

Jerry stammers. "Well, I guess you feel kinda good, you know, relaxed like. Mellow, you know. And then you get dizzy, maybe a headache."

"Okay, what else?" the trainer asks. "Let's say you go past feeling good to feeling dizzy and you keep right on drinking. What happens next?"

"You get sick and pass out."

"Okay, what you have just described is a toxic reaction to overexposure to a chemical called ethyl alcohol. Most people call it getting drunk."

"Ethyl alcohol? Like the stuff in gasoline?"

"Yep."

"No wonder my head hurts on Sunday morning."

"And why does the ethyl alcohol make you act funny and feel funny and get sick?" the trainer asks.

"Because it goes to your head."

"Right, but how?"

"In your blood."

"So let me ask you this: Is there any time, other than when you're drinking alcohol, that you have experienced sensations similar to what you feel when you are drunk?"

One man says: "I know sometimes after I've been welding galvanized, I feel real sick to my stomach for hours."

Another man adds: "When I have to use a lot of trichlor, you know, degreasing parts, I get a kind of heavy feeling in my head."

Jerry chimes in: "We all know what it's like painting outside of the paint booth."

"Okay, let's find out why breathing those chemical vapors makes you feel the same as drinking too many beers." The trainer passes out selected Material Safety Data Sheets for the materials the men have mentioned.

KEEP IT SIMPLE

Simplicity should be your by-word and slogan. Remember, changing behavior through safety and environmental training is a cumulative process. The first step is creating a *desire for it* in your students. Once this common ground is established, you move ahead one small goal at a time. Don't expect your students to absorb everything at once or to change overnight.

For each training class or safety meeting, set a goal of one or two practical concepts that the students can take back to work and apply right then, that same day. For purposes of Hazard Communication training, such concepts might include: "Safe storage and handling of flammables," "Why corrosives are loners," "How to recognize overexposure," and "Reading labels can save your life." For Lockout/Tagout training, the focus might be limited to: "The differences between locks and tags," "Safety during routine maintenance," and "How to isolate equipment from energy." For respirator training, some class themes might include: "Why is a respirator like insurance?" "How to clean and store your respirator," and "What can a respirator do for your lungs?" For RCRA Hazardous Waste training, topics might include: "Managing waste to minimize it," "How hazardous waste is disposed of," and "How the waste we generate impacts our environment."

Use one or two concepts to form the backbone of each class. Don't get overanxious and worry that students must learn everything at once. Instead, remind yourself that you will have another chance to talk with them. Each class is a beginning. Something to *build on*.

Your goal must be to foster a desire, to initiate an awareness. It is far better to get only one point across in each class than to turn students off by overwhelming them.

Is it okay to make it fun? Yes. Safety training doesn't have to be synonymous with suffering. Remember the boring speaker paraphrased on page 194? There is nothing simple, nothing accessible, and certainly nothing fun about his training style. As a consequence, there is probably nothing about the classes that the students remember, except how much they dread them.

What can you do to make your classes more fun? Aside from keeping your classes tightly focused and building them around employee experiences, you can be as creative as your mind allows in designing your classroom format. Loosen up. Surprise your students. Don't be afraid to be nontraditional, to break away from the videos-and-comic-books rut into which training classes often fall. Experiment. What, after all, do you have to lose?

Use Music

For instance, for a class on accident prevention you might open the session by playing Aretha Franklin's song "Think," a wonderful anthem to the value of thinking before acting. There are countless other songs of all genres that can be effective and entertaining in a training class.

Use Games and Competition

Break your class into teams and have them compete to label a chemical container or to find information on a Material Safety Data Sheet. Have them conduct a "scavenger hunt" to locate energy-isolating devices or transcribe information from labels. Develop your own crossword puzzles or matching games, where employees are challenged to link chemical concepts or label terms with the actual products or situations they encounter on the job. Tailor the games to your employees' skills and needs. If they have poor writing skills, don't emphasize written games. People like to compete. Regardless of the makeup of your class, whether you're talking with managers, chemists, production workers, or janitors, friendly competition will not only enliven your class; it will make the subject matter more memorable. If Department A beats Department B in a contest regarding chemical labeling, the labels themselves take on a new image—that of a source of pride out in the workplace.

Use Reenactments

There is a safest and best way of doing everything at your company. Using classroom reenactments is one way of bringing safety out of the cold realm of abstract theory and placing it in the corporeal context of daily worklife. Recruit a student to demonstrate putting on a respirator, and have the rest of the class critique him or her. Have two employees turn a 55-gallon drum on its side or loosen bungs to take a sample, and have the rest of the class discuss what they do correctly and incorrectly. Encourage your students to come up with their own reenactments, and then build the class discussion around them.

Use the Workplace Itself

One of the most overlooked tools available to the trainer is the workplace itself—the real thing. Instead of relying upon videos of *other people's* companies, why not use your own? Take the class out of the breakroom or conference room and move it to the production floor. Instead of *reenacting* work tasks, have your students do the real thing—under the critical eyes of their coworkers.

As a twist on the status quo, allow your *students to play inspector.* Let's say that the topic of the day is safe chemical handling. Give each of your students a pad and pencil and have them go out into the workarea and note what they observe. Are tops and lids on chemical containers? Are drums grounded? Are containers labeled? Are flammables and combustibles isolated from ignition sources? Are incompatible chemicals stored separately? After they have had a chance to tour the workplace, get everyone back together and discuss their findings. This is a highly effective method for linking safety with operating procedures, or what's really going on out on the production floor. This simple yet novel approach can work with any classroom topic.

Photographs of the workplace and of employees at work can serve as the simple, and often amusing, basis of a training class. People like to see themselves and their coworkers in photos. Photos can be analyzed for evidence of safe and unsafe procedures in much the same way that reenactments and live workplace activities can be analyzed. In fact, often employees can "see" things in photos that they are blind to when actually standing in the workarea. Photos provide distance that allows employees to see themselves and their workarea more objectively.

It is probably impossible to err on the side of simplicity where training is concerned. If you get an idea, don't dismiss it because you think it might be too simple or silly, or because you're concerned that it has never been done before. *Try it.* Your students will let you know soon enough if it doesn't work.

Nothing replaces the team spirit and positive attitudes fostered while people are challenged both *to learn* and *to enjoy* themselves. This is a simple concept. Let your training be no more complex.

WHOSE LIFE IS IT, ANYWAY?

Tell me, and I'll forget.
Show me, and I may not remember.
Involve me, and I'll understand.

Native American saying

It is an unfortunate but common sentiment among employees that their employer doesn't care about them. Counter this attitude with a challenge: Tell them that you wouldn't be there if you didn't care, but that caring is a two-way street. You can care enough to listen to them, provide them information, and get them the equipment they need, but they must care about their own bodies enough to use it. After all, it's their life.

Establish that the purpose of training is to provide them with the information they need to protect themselves from the hazards of the workplace. The information can help them make educated decisions about how to do their jobs safely. However, the decision to improve the quality of their worklife and to protect their health by following safe operating procedures, reading labels, and wearing protective equipment is strictly up to them. You can't do that part for them.

Self-respect is central to the learning process. The trainer who overlooks student attitudes toward themselves and their jobs might as well be conducting classes in a foreign language. Remember, if students don't value their own lives, they won't value your Hazard Communication or Lockout/Tagout training. Let building self-respect form the core of all you say and do.

Autonomy Versus Fate

"They just don't care about safety and health." This is a common management complaint. But how valid is it? Most people care about something: spouse, kids, house, car, keeping their job, staying alive. *Something* matters. What you must do as their trainer is find out what matters to them and build a bridge *connecting that care with safety and health.*

Some employees will insist on taking a maddeningly cavalier attitude toward their health. They'll tell you that they might get hit by a truck tomorrow, that they might die of any number of causes. They'll ask why they should suffer the hassles of changing their behavior at work to protect themselves from other equally uncontrollable hazards? They'll tell you they'd prefer to take their chances. When your time's up, it's up. No use fighting.

They'll tell you that they are too young to change, too young to worry about wearing protective equipment and using machine guards. They're strong, tough. Nothing bad is going to happen to them. Or they'll tell you they're too old to change. If something bad was going to happen to them, it would have happened ten years ago. Or, perhaps, it has already happened and they just don't know it yet. Either way, no use crying over spilt milk now.

For a myriad of reasons, employees often don't understand that contracting a disease from occupational exposure to chemicals or losing a finger in a machine because a safety device was by-passed do not fall into the same category of random events as being hit by a drunk driver or being mugged in a convenience store. Occupational accidents and illnesses are, in large part, not a matter of luck or fate.

They are a matter of *autonomy and self-control.*

But employees aren't taught to think this way about their jobs. They are taught to do what they're told and get paid. For all their posturing about how tough and impervious they are, when you get down to the nitty gritty, most employees *expect* to suffer in their work and because of it. They expect to be uncomfortable, to get hurt, and to experience negative effects from working with chemicals. Ill-equipped or unaware that they are capable of preventing these problems, they knuckle under and bear them, convinced they have no choice. The problem is that inevitably, over time, they come to resent the hell out of it.

Our current litigious climate only encourages this resentment. The system doesn't encourage employees to take responsibility for themselves, to work safer and smarter. Instead, it says: *Your safety and health isn't your responsibility, here, sue your employer for not taking care of you.* Neither does it encourage employers to demand that their employees respect and thus protect themselves. Instead, it says: *Hey, you complied with the law, all the documentation is right here, if your employees choose not to work safely, that's their problem.*

Our legal system encourages both sides to blame one another so that neither has to change. And everybody loses.

If you are interested in cutting the losses at your company, you have to start by letting each one of your employees know that working safely is their personal responsibility. Doing things safely may not always be easy, it takes discipline and commitment, but, in the long run, adopting a personal policy of working safely is the best kind of insurance they can invest in. Because (and everybody needs to be reminded of this), *work is not separate from real life; it's essential to it.* Your employees will spend at least half of their lives working. To apply a separate, and often lesser, standard of safety and health to work than to other parts of their lives is absurd.

Let's take chemical exposure as an example:

Your employees can tell you about the chronic or long-term effects of drinking too much alcohol or smoking. They can also describe some of the acute or immediate physical sensations experienced while drinking a beer or smoking a cigarette. Knowing these risks, some people have decided never to smoke or drink. Some cut back on either or both. And others smoke and drink as if they never heard risk was involved. If you ask this last group why they haven't changed, they'll tell you it's because they don't want to, they can't, or they just enjoy it too much. In their perception, the benefits (in pleasure or comfort) outweigh the negatives (in risk of disease). Fine. It's logical. You may not agree with it. But they've made their choice.

Your employees can describe certain work activities where they experience sensations very similar to those experienced while drinking or smoking. Where they get high or dizzy, where they become nauseated, or get a headache, dry throat, burning sensations in the chest, or watering eyes. Some employees consider sensations such as these to be *part of the job.* They don't like them, but they don't worry about them either. Consequently, the acute sensations themselves, and even the threat of possible long-term or chronic health effects resulting from exposure to the substances that caused the sensations, is not enough to shake them up. It is certainly not enough to prompt them to take steps to stop experiencing the sensations. If you ask them why they don't want to wear a respirator or gloves or take other precautions to protect themselves from acute and chronic health effects, they'll tell you it's because they *can't.* "I can't work with that stuff on." "I can't work that way." "I've always done it this way."

And that's when you ask them: "You mean to tell me it's not because you *want* to feel those sensations? It's not because you enjoy it and it brings you pleasure? You mean to tell me that experiencing these sensations is unpleasant and has no benefits for you at all? And yet you go on suffering with them because you won't change a procedure? Are you really telling me that you are willing to run the risk of affecting your health years down the road because you won't learn to work in a respirator and gloves?"

About that time, you'll have everyone's attention. Tell them to take out that insurance policy, by changing their behavior. Not for you, not for the company, but *for themselves.*

Pulling the Heartstrings

Sometimes, in your efforts to initiate behavior change, it may be necessary to switch the focus to that of the health and well-being of your employees' families. This approach can be effective in several contexts. As mentioned earlier in the chapter, everyone needs a motivation for change. For some people, reducing the risk of their own untimely demise is enough motivation to prompt them to overhaul their behavior. As soon as they are informed of the risks of any given activity and determine that those risks outweigh any benefits, presto! They change.

Other nuts are harder to crack. For them you might need to find motivation in their families. Set up a pros and cons sheet for any safety procedure that they are particularly stubborn about. There are always more cons than pros to doing a job unsafely. Always ask for the pros first. When seen in black and white, the ludicrousness of the rationale for working unsafely is clearly illuminated. What are the benefits of operating a grinder without safety glasses? *Saves time because you don't have to put the glasses on. Allows you to work faster because you can see the part better. Is more comfortable.* How do these stand up against the possible negative results, for a father of four young children, if he damages or loses an eye?

Perhaps your employees have a negative attitude about certain procedures you are training them to follow for the management of hazardous waste or the cleaning up of spills. Perhaps they view the procedures or paperwork as pointless government intervention that has no connection to their work or their life. Again, this is a perfect time to bring in the family. Draw the connection between their activities at work and the impact it has on the community, and thus the lives of their spouses and children.

Why must hazardous waste be managed and disposed in accordance with strict rules and accompanied by special labels and paperwork? To keep it from being unsafely dumped on roadsides and in fields where it can contaminate soil and water. Why must chemical spills be cleaned up in a timely manner and disposed of properly? To limit exposure of the chemical to air and to limit the possibility of the spill reaching groundwater. In other words, to reduce contamination of the environment their families inhabit.

As employees come to understand that their activities at work affect their lives and those of their families outside work, that working in a safe and healthy manner has a long-range impact, actually giving them a little autonomy, a little influence over their future, they will begin to change their behavior. And when you see the first changes occurring, that's when you know your training is working *for you.* For the company. For the well-being of *each individual,* without whose impact and influence your organization would achieve but a fraction of its potential quality.

Postscript

*The true revolutionary is guided by
a great feeling of love.*

Che Guevara

I grew up on an industrial site in Memphis. But in June 1974, the year my parents purchased the site, it wasn't yet industrial. It was just a field full of black-eyed susans, rabbits, and bullfrogs.

That first summer, my five-year-old brother and I accompanied my mother door to door in the neighborhood adjacent to the property. When mother knocked on the doors of the small box-like houses that lined the street that led to our field, she knocked as a mother asking other mothers for help. Her problem? Someone, probably neighborhood kids, was systematically tearing down the property line fence and smashing the cinder blocks of the shed my father struggled each day to build.

Every morning, the previous day's work was demolished. Every night, my father threatened violence. Mother was desperate for relief.

"We don't have a lot of money," she would confess, standing on someone's porch, one hand on my brother's shoulder, one hand on mine. "We're trying to build a business here. A business that will create jobs for the people of this neighborhood. If you know who is doing this, please, will you tell them to stop?"

Her approach was straightforward. Her plea, honest. The only catch was that when the women of the neighborhood opened their doors, all they saw was another woman accusing their children. As soon as they realized what she wanted, their hearts and mouths snapped shut as quickly as their doors. And who could blame them?

What my parents experienced that first summer and continued to experienced for nearly 15 years was a *failure to communicate*. It was a failure born not out of hatred or malice and certainly not out of lack of effort, but out of an inability to put themselves in the other person's shoes. An inability to see things through other's eyes.

Over the years, as my parents operated their small solvent recycling company, I spent summers "at the plant," which was still mainly an outdoor facility, a gravel-covered field sprinkled with above-ground storage tanks, a Quonset hut, a portable building, and several lean-tos. As I hand-pumped small solvent orders, painted tanks, or otherwise messed around, I watched my parents' employees, many of whom were in fact from the neighborhood and most of whom were poor and undereducated.

Later, in thinking back over that time, I realized that my parents were often played for fools by their employees. Despite their Trojan efforts to treat everyone fairly, they were almost continually taken advantage of. Their workers were undisciplined. There was little evidence of care or pride.

Why?

It certainly wasn't because they thought themselves far superior to their employees. Quite the contrary! Their personal commitment to race and class blindness never allowed them to *recognize* a difference between themselves and the people who worked for them. They assumed that, when given the opportunity, their poor, undereducated workers would perceive things (things like right and wrong, work, trust, freedom, discipline, goal-setting, and self-improvement) exactly as they themselves perceived them. They repeatedly were shocked that their employees, to whom they had given so much opportunity, just kept throwing it back in their faces.

But why did this happen? Just bad luck? An unfortunate set of circumstances exacerbated by the temper of the times? I don't think so. I think what my parents experienced is based on a much more universal truth. My parents, their staunch belief in the Golden Rule notwithstanding, failed as managers because they didn't respect the people they were trying to manage.

Shocked? Just think about it a minute. What does respect mean? *To show esteem for. To show consideration for. To relate to.*

Is it respectful of someone to assume, blanketly, that they are just like you? That they view life and work with the same expectation?

Is it respectful to assume that, since *you* want them to have responsibilities and goals, that *they* want those burdens?

Is it respectful of someone to assume that, when exposed to your views, they can and will rush to adopt them as their own?

No, it's not respectful. It's egotistic.

By contrast, the most respectful approach that a manager can take is to forget for a time his or her own attitudes and prejudices and "the way I would act if that was my job" and open his or her mind to the employee's perceptions. Find out where they are coming from, and then go out to meet them on some common ground—some place where experience truly is shared and "difference" is not so threateningly apparent. When a manager acts from respect, he or she is seeing the employees as they really are, all strength and weakness, difference and sameness— A perception that will allow him or her to provide effective leadership and discipline.

When a manager acts from ego, no matter how well-intentioned, he or she is seeing the employees as reflections of him or herself. Employees are quick to pick up on this and to use it to their advantage, for they know that, under such circumstances, their boss' vision is about as clear as an image in a fun house mirror.

As a trainer, the closed faces of the neighborhood women of my childhood have greeted me again and again in the faces of men and women of all ages, races, nationalities, and learning abilities. I have met them in conference rooms, lunch rooms, break rooms, warehouses, in the middle of high-tech plants, and in grungy portable buildings and lean-tos.

When I have stood in front of them I have, on occasion, looked down from the high ground of knowledge and been self-satisfied. At other times I have looked across a chasm of differentness and been terrified. And, ultimately, I have hacked my way through the underbrush of bias to a clearing called *common ground.*

In the process I have discovered that people really are easy to love. Easy, that is, as soon as you accept that their thinking or acting or talking like you is not a requirement.

Appendix

You Are Not Alone

Sooner or later, every trainer needs help—resources, references, back-up, support, even a shoulder to cry on. You need associations, agencies, and colleagues who will validate your decisions and share your successes and failures. You need guidebooks to which you can go for answers that won't be questioned later on.

You need to be able to verify your interpretation of a regulation, the proper spill response for a chemical, how a waste should be disposed of, or what type of gloves will provide appropriate protection for a certain job. You need fresh ideas for training classes, such as a new video, a game, or an incentive program. You need regulatory updates and information on new trends and technologies. Perhaps you just need a friend.

The following pages contain a few of the places you can turn to for this kind of help. These lists are by no means comprehensive. They provide a place *to begin your search* for assistance. They do not define the boundaries of your search or limit where it might take you.

A disclaimer (and a word of praise) regarding associations and training schools: Although millions of dollars are spent each year by professionals seeking memberships in associations, the sole purpose of which is to have something to put on their resume, I am a fan of associations because they have helped me grow as a trainer. But the only reason they have helped me is because I have been an *active member*. Active membership means going to conferences sponsored by the association, attending local chapter meetings, participating on committees, and writing articles to be published in the association's newsletter or magazine. All of these things take time and money, which for me, as a small business owner, has translated into brute sacrifice. Has it been worth it? *Absolutely.* What I have gained in terms of ideas and information exchange, confidence, professional stature, and comradeship is immeasurable.

Are you feeling down in the dumps and discouraged? Are you less than confident about your ability to be an effective trainer? Rather than a new job or a vacation, what you may need most is to know that you are a member of a community of like-minded people—people who get up every morning and face the same kinds of trials and tribulations. You need to know you're not alone.

I can't guarantee that the associations listed here will do this for you. Pick one or two that seem to address your needs and particular expertise and call them. For an end to loneliness, it's worth a try.

As for training schools, there are no guarantees either. The quality of train-the-trainer courses and seminars currently available in the marketplace is as varied as the subject matter they cover. You can waste a lot of money and time attending courses that deliver very little in practical information that you can apply at your workplace. On the other hand, some training courses offer concentrated, focused information that can lead you, like torchlight, out of the darkness of regulatory confusion.

Is selecting a quality course merely a matter of blind luck? Since there is no quality assurance program or national clearinghouse for training courses, to some extent, yes, luck does figure into it. However, there are a couple of key rules that will give you a leg up in assessing the probable usefulness of a course.

First, *less is more*. Beware of seminars and courses that are billed as overviews. These courses play on our desire to get a bargain. "Wow," we think as we look at the literature for the course, "everything I need to know to comply with OSHA and EPA regulations in only six hours. What a deal!" Wrong. Of much more value would be a six-hour course focused on a specific regulation (such as Hazardous Waste Management or Emergency Response) or area of compliance (such as training or recordkeeping). Comprehensive courses tend to be accompanied by huge notebooks of information. Sometimes these books contain new data, forms, tables, and flowcharts you can use to enhance your compliance back at work. More often, however, they contain only copies of the regulations. Most of which you probably already have. A course focused on one or two objectives will always be of greater value than a comprehensive overview.

Second, *if it sounds too good to be true, it is*. Beware of courses that claim miracle solutions. "Learn how you can meet all your OSHA compliance obligations in just one hour!" Compliance is work. Compliance takes time. Training employees so that procedures change and the workplace improves is not an overnight endeavor. If a training course leaves out the work and the time that must be invested in order for training to succeed, there are probably a lot of other things it leaves out, too.

Continuing education is vital to your effectiveness as a trainer. You can't afford not to stay "up" on the regulations that impact your workplace; neither can you afford not to be exposed to new approaches to classroom training and to new training materials.

Training courses and seminars can do these things for you. But you have to choose them carefully, which, like everything else about the compliance process, takes personal effort.

ASSOCIATIONS

Air and Waste Management Association, P.O. Box 2861, Pittsburgh, PA 15230. Phone: (412)232-3444. Provides technical and managerial information about environmental issues. Offers training courses and technical and managerial information exchange network. Members receive a yearly subscription to the *Journal of Air and Waste Management Association,* discounts on registration fees for meetings and courses, discounts on the association's publications and videos, and a subscription to the bimonthly newsletter *News & Views.* Annual dues: $80.

American Institute of Hazardous Materials Management, 900 Isom Road, Suite 103, San Antonio, TX 78216-4102. Phone: (800)729-6742. An educational institute and membership association, the purpose of which is to promote the effective management of the risks of hazardous materials, including risks associated with the regulated status of and public antipathy toward hazardous materials. Members receive a certificate of membership, the Institute newsletter, *On Managing Hazardous Materials,* and discounted rates on educational seminars and courses offered by the institute. Educational activities include seminars and courses for managers and executives; training programs for those who handle or respond to hazardous materials; conferences; development and presentation of educational materials and programs for business, industry, institutions, and the public; and other educational activities. Annual dues: $75.00.

American Society of Safety Engineers, 1800 East Oakton Street, Des Plaines, Illinois 60018-2187. Phone: (708)692-4121. **Founded:** 1911. **Members:** 25,000. Training seminars, publications, audio-visual training courses, safety seminars, professional development conference and exposition. ASSE also provides the following services to members: Professional development conference and exhibition, national education and training seminars, technical publications and audiovisual materials, compensation and benefits survey data, JobLine and membership directory, annual leadership conference, *Professional Safety* magazine, and *Society Update* newsletter. New member dues: $115.00.

American Society of Testing and Materials, 1916 Race Street, Philadelphia, PA 19103-1187. Phone: (215)299-5585. ASTM is a technical publisher. Develops and publishes standard specifications, tests, practices, guides, and definitions for materials, products, systems, and services. Publishes books containing reports on new testing techniques and their possible applications. Annual dues: varies.

International Fire Service Training Association, Oklahoma State University, Engineering Extension, 512 Engineering North, Stillwater, OK 74078-0532. Publications and manuals for fire service training. Has fire service training department. Annual dues: varies.

National Environmental Training Association, 8687 Via De Ventura, Suite 214, Scottsdale, AZ 85258. Phone: (602)956-6099. Charles L. Richardson, Exec. Dir. **Founded:** 1977. **Members:** 100. Professional environmental trainers organized to promote better operation of pollution control facilities by means of personnel development. Organizes and delivers training in the fields of water supply, wastewater, air and noise pollution control, and solid and hazardous waste. Purposes are to encourage communication among individual trainers, training institutions, and government agencies; to promote environmental personnel training and education; to set minimum standards for training and education programs; and to encourage reciprocity between states and training institutions on evaluation and acceptance of transfer credits. Sponsors research programs and seminars; bestows Environmental Education and Trainer-of-the-Year awards. Operates information center. Conducts a national certification program for environmental trainers. Publishes *Membership Directory,* annually and *NetaNews,* bimonthly.

Covers association activities and efforts to inform Congress and the Environmental Protection Agency of environmental training issues. Includes calendar of events. Price of publications is included in $50 annual membership or $12/year for nonmembers.

National Fire Protection Association, 1 Batterymarch Park, P.O. Box 9101, Quincy, MA 02269-9101. Publishes a variety of materials on fire codes and standards, hazardous materials, fire protection, fire service, and fire-safety education. Annual dues: varies.

National Safety Council, 444 North Michigan Avenue, Chicago, Illinois 60611. Phone: (800)621-7619. Publishes a huge variety of magazines, newsletters, journals, and audio and video materials on a variety of safety topics relating both on and off the job. Professional resources available to assist in developing and maintaining effective safety and loss control programs. The Loss Control Consulting Group can offer assistance in the following areas: safety program review/needs assessment; environmental/safety and health audits; custom program/materials development; training program development/presentation; ergonomic surveys/training; behavioral analysis; safety management resource; fleet safety program development/implementation; industrial hygiene survey. Annual dues: varies.

SOURCE BOOKS AND ASSISTANCE GUIDES
Emergency/Spill/Fire Response

Emergency Handling of Hazardous Materials in Surface Transportation
Bureau of Explosives

Association of American Railroads
1920 L. Street, N.W.
Washington, D.C. 20036

Emergency Action Guides
Bureau of Explosives
Association of American Railroads
1920 L. Street, N.W.
Washington, D.C. 20036

CHRIS Manual
Superintendent of Documents
Government Printing Office
Washington, D.C. 20402-9325

1984 Emergency Response Guidebook: Guidebook for Hazardous Materials
U.S. Department of Transportation
Materials Transportation Bureau
Attention: DMT-11
Washington, D.C. 20590

Fire Protection Guide on Hazardous Materials
National Fire Protection Agency, (NFPA)
470 Atlantic Avenue
Boston, MA 02210

Fire Officer's Guide to Emergency Action
NFPA

Fire Protection Handbook
NFPA

Fire Officer's Guide to Disaster Control
NFPA

Pre-Planning and Guidelines for Handling Agricultural Chemical Fires
National Agricultural Chemical Association
1155 15th Street, N.W.
Washington, D.C. 2005

Hazardous Waste Management Guide
J. J. Keller & Associates, Inc.
145 W. Wisconsin Avenue
Neenah, WI 54956, (414)722-2848

Hazardous Materials Guide
J. J. Keller & Associates, Inc.
145 W. Wisconsin Avenue
Neenah, WI 54956, (414)722-2848

Industrial Toxicology, Chemistry of Hazardous Materials
Prentice-Hall, Inc.
Englewood Cliffs, NJ

Fire and Flammability Handbook
Van Nostrand Reinhold
115 Fifth Ave.
New York, NY 10003

Advanced First Aid and Emergency Care
American Red Cross

First Aid Manual for Chemical Accidents
Van Nostrand Reinhold
115 Fifth Ave.
New York, NY 10003

Personal Protective Equipment

Guidelines for the Selection of Chemical Protective Clothing, 2nd Edition
American Conference of Governmental Industrial Hygienists, Inc., (ACGIH)
6500 Glenway Avenue, Bldg. D-7
(513)661-7881

Protecting Personnel at Hazardous Waste Sites
Butterworth Publishers
80 Montvale Avenue
Stoneham, MA 02180

The Merck Index
Merck & Company, Inc.
Rahway, NJ

214   Safety and Environmental Training

Personal Protective Equipment for Hazardous Materials Incidents
NIOSH, ACGIH

6500 Glenway Avenue, Bldg. D-7
Cincinnati, OH 45211

General Hazardous Substance

Gas Data Book, 6th Edition
Matheson Gas Products
Secaucus, NJ 07094

Pocket Guide to Chemical Hazards
NIOSH/OSHA
Superintendent of Documents
Government Printing Office
Washington, D.C. 20402

Handbook of Toxic and Hazardous Chemicals and Carcinogens
Noyes Publications
Mill Road
Park Ridge, NJ 07656

Condensed Chemical Dictionary
Van Nostrand Reinhold
135 West 50th Street,
   New York, NY 10020

Safe Storage of Laboratory Chemicals
John W. Wiley & Sons, Inc.
New York, NY

Standard Operating Safety Guide
EPA Office of Emergency and Remedial Response
Hazardous Response Support Division
Environmental Response Team
November 1984

Labeling

Handbook of Chemical Industry Labeling
Noyes Publications
Mill Road, Park Ridge, NJ

Regulations

Occupational Safety and Health Reporter
Bureau of National Affairs
1231 25th Street, N.W.
Washington, D.C. 20037

29 Code of Federal Regulations Part 1910 (OSHA)
Superintendent of Documents
Government Printing Office
Washington, D.C. 20402

40 Code of Federal Regulations(RCRA)
Superintendent of Documents

Government Printing Office
Washington, D.C. 20402

49 Code of Federal Regulations Parts 170-199 (DOT)
Superintendent of Documents
Government Printing Office
Washington, D.C. 20402

Environment Reporter
Bureau of National Affairs
1231 25th Street, N.W.
Washington, D.C. 20037

Chemical Regulation Reporter
Bureau of National Affairs
1231 25th Street, N.W.
Washington, D.C. 20037

ASSISTANCE ORGANIZATIONS

Regulations

Oil and Hazardous Material Technical Assistance Data System,
EPA

Oil and Special Materials Control Division
Office of Water Program Operation
Washington, D.C. 20460

Emergency Fire/Spill

Chemical Transportation Emergency Center (CHEMTREC)
Manufacturing Chemists Association
1825 Connecticut Avenue, N.W.
Washington, D.C. 20009

(800)424-9300
CHEMTREC Information Number
(800)CMA-8200, D.C. 887-1315
Alaska (202)887-1315

TRAINING SCHOOLS (TRAIN-THE-TRAINER)

Asbestos Institute, 1130 Sherbrooke Street, W., Suite 410, Montreal, PQ, Canada H3A 2M8. Phone: (514)844-3956. Organizes and participates in training programs and information sessions; sponsors seminars. Maintains documentation and reference bank on asbestos. Publishes *ASBESTOS* (in English, French, and Spanish), quarterly. Also publishes bulletin and brochures. Holds periodic conferences.

Association of University Programs in Occupational Health and Safety, c/o Dr. Roy E. Albert, Institute of Environmental Health, University of Cincinnati, 3223 Eden Avenue, Cincinnati, OH 45267. Phone: (513)558-5701.
Founded: 1977. **Members:** 14. Universities offering graduate and continuing education for occupational safety and health professionals. Provides a forum for the exchange of information among members on graduate training in occupational medicine, occupational health nursing, industrial hygiene, and industrial safety engineering. Works in conjunction with the National Institute for Occupational Safety and Health (NIOSH) to facilitate the operation of training programs. Compiles statistics.

Georgia Tech Research Institute/ESTL, Atlanta, GA 30332-0385 Phone: (404)894-3806. Provides up-to-date technical training and information. Offers more than 120 courses annually on a wide variety of environmental subjects. Subject areas include industrial hygiene principles and practices, environmental auditing, asbestos identification, abatement and management, hazardous material control and emergency response, underground storage tanks, lead-based paint abatement, hazardous waste management, OSHA Hazard

Communication, wetlands determination, ergonomics and safety, indoor air quality, environmental engineering, and industrial processes.

Institute of Hazardous Materials Management, 5010 A Nicholson Lane, Rockville, MD 20852. Phone: (301)469-8969. Offers certification program in hazardous materials management. Hazardous Materials Management Certification Program has the following objectives: to provide credentialed recognition to those professionals engaged in the management and engineering control of hazardous materials who have attained the required level of education, experience, and competence; to foster continued professional development of Certified Hazardous Materials Managers through continuing education, peer group interaction, and technological stimulation; to facilitate the transfer of knowledge and experience among professionals; to provide government, industry, and academia with a mechanism for identifying hazardous materials management professionals who have fulfilled the requirement for certification by a professional peer group. Offers membership to Academy of Certified Hazardous Materials Managers, which provides a mechanism for further professional development and for exchange of ideas and information to advance the practice of hazardous materials management. Annual Academy dues: $15.00.

Louisiana State University, Environmental Occupational and Safety Training Office, 177 Pleasant Hall, Baton Rouge, LA 70803. Phone: (800)256-6948.

McGill University, School of Occupational Health, 1130 Pine Avenue W., Montreal, PQ, Canada H3A 1A3, Dr. Gilles Theriault, Director. Phone: (514)398-7435. Provides continuing education programs in occupational health through the Educational Research Center. Sponsors a work environment health seminar series.

Midwest Center for Occupational Health and Safety, Box 197 Mayo Memorial Building, 420 Delaware Street, S.E., Minneapolis, MN 55455. Phone: (612)221-3992.

National Environmental Training Association, 2930 East Camelback Road, Suite #185, Phoenix, AZ 85016-4412. Phone: (602)956-6099. Train-the-Trainer Workshop Series. Individual workshops in hazardous materials regulatory compliance, designing effective environmental training, delivering effective environmental training, and managing environmental training. Certification is available in asbestos abatement, Hazard Communication Standard, hazardous materials and waste management, occupational health and safety, transportation of hazardous materials and waste, wastewater treatment, and water treatment.

National Fire Academy, Emergency Management Institute, Office of Admissions, 16825 South Seton Avenue, Emmitsburgh, MD 21727. Phone: (301)447-1035. Offers resident and non-resident courses under the following curricula: Community preparedness/exercise programs, professional development, mitigation and natural hazards, radiologic, hazardous materials, and national security. Other training activities include an emergency management workshop.

Oklahoma State University, Engineering Extension, 512 Engineering North, Stillwater, OK 74078-0532. Annual Environmental Workshop featuring the following courses: hazardous materials management certification, CHMM/REM certification, OSHA HAZWOPER training, real estate property assessor, building an effective health and safety program, the nuts and bolts of industrial hygiene, confined space-lockout/tagout, and certified environmental auditor.

Queens University at Kingston, Occupational Health and Safety Resource Centre, Abramsky Hall, 2nd Floor, Kingston, ON, Canada K7L 3N6. Phone: (613)545-2909. Sponsors industrial audiometry and hearing conservation, spirometry workshop, refresher course audiometry (annually), and short course series. Provides consulting for occupational health, environmental hazards, and health contract research.

Roane State Community College, Waste Management Training Center, 728 Emory Valley Road, Oak Ridge, TN 37830. Phone: (800)343-9104. Provides programs for occupational health training, hazardous materials management, asbestos management, and related areas of practice. Trains and certifies personnel in a variety of waste management fields. A leader in OSHA and asbestos training.

Texas A & M University, Occupational Health and Safety Institute, Department of Industrial Engineering, College Station, TX 77843. Phone: (409)845-5531. Offers continuing education courses.

University of California, Davis Division of Occupational and Environmental Medicine, Davis, CA 95616. Phone: (916)752-4256. Holds an annual Northern California Occupational Health Symposium, open to occupational medicine practitioners, interns, and representatives from the industry.

University of South Alabama, Center for Emergency Response Training, USA Brookley Center, 245 A Club Manor Drive, Mobile, AL 36615. Phone: (205)431-6527. Offers comprehensive training in all areas of HAZMAT related response and also industrial fire brigade training. Courses include hazard awareness and response management, managing the hazardous materials incident, treatment storage and disposal facility training, railroad emergency actions, hazardous materials tactics, hazardous materials refresher course.

University of California, Northern California Occupational Health Center, San Francisco General Medical Center, Building 30, Fifth Floor, 1001 Potrero Avenue, San Francisco, CA 94110. Phone: (415)821-5200. Trains occupational physicians, nurses, toxicologists, epidemiologists, and industrial hygienists. Also conducts continuing education and outreach programs for occupational health professionals and workers.

University of Findlay, Emergency Response Training Center, 1000 North Main Street, Findlay, OH 45840. Phone: (419)424-4647. Provides education, training, and information transfer relative to hazardous and potentially hazardous materials, hazardous waste, and environmental concerns of any nature. Offers one to five day workshops covering such topics as OSHA safety training, emergency response, confined space entry and rescue, property assessment, RCRA generator liability, waste minimization and recycling. Offers academic degree programs.

University of Pittsburgh, Center for Hazardous Materials Research, 320 William Pitt Way, Pittsburgh, PA 15238. Phone: (800)334-2467. The training facility is outfitted for transportation, hazardous waste operations, in-plant emergency response, and community emergency response training exercises.

CHMR currently offers 19 courses in four categories: hazardous waste operations, hazardous materials emergency response, occupational health and safety, and environmental compliance. The courses are designed to meet the training requirements of OSHA, RCRA, DOT, TSCA, and NFPA and have been qualified by the American Industrial Hygienists Association for Continuing Education Credits.

University of Utah, Rocky Mountain Center for Occupational and Environmental Health, Building 512, Salt Lake City, UT 84112. Phone: (801)581-8719. Offers continuing education programs in industrial hygiene, occupational medicine, occupational health nursing, and occupational safety and ergonomics (over 1,000 attendees per year). Provides consultation and service to federal, state, and local governments, industry, and other organizations.

USEPA, Emergency Response Training Program, 26 West Martin Luther King Drive, Cincinnati, OH 45268. Phone: (513)568-7537. Courses offered in Regions I-X include treatment technologies for Superfund, air surveillance for hazardous materials, hazardous materials incident response operations, risk assessment guidance for Superfund, introduction to groundwater investigations, safety and health decision-making for managers, sampling for hazardous materials, radiation safety at Superfund sites, emergency response to hazardous material incidents, advanced air sampling for hazardous materials, and removal cost management system.

TOXICOLOGY AND EXPOSURE LIMITS

American Conference of Governmental Industrial Hygienists, Building D-7, 6500 Glenway Avenue, Cincinnati, OH 45211-4438. Phone: (512)661-7881. Publications: 1990-1991 *Threshold Limit Values for Chemical Substances and Physical Agents and Biological Exposure Indices, ACGIH Transactions, Applied Occupational and Environmental Hygiene.*

National Toxicology Program, Public Health Service, Public Information Office, P.O. Box 12233, MD B2-04, Research Triangle Park, NC 27709. Phone: (919)541-3991. *Fifth Annual Report on Carcinogens* (1989), *Chemical Status Reports*, Long-Term Technical Reports, *Short-Term Toxicity Study Reports.*

International Agency for Research on Cancer, 150 cours Albert Thomas, F-69372 Lyon Cedex 08, France. Phone: 7 72738485. United Nations Organization. Publications: *Biennial Report, Directory of Ongoing Research in Cancer Epidemiology, Monographs.*

American Medical Association, 515 North State Street, Chicago, IL 60610. Phone: (312)464-5000.

National Institute for Occupational Safety and Health, 4676 Columbia Parkway, Cincinnati, OH 45226. Phone: (513)533-8236. *Pocket Guide to Chemical Hazards.*

PERIODICALS

Business and Health
Topics: Accident prevention; occupational safety; safety training; industrial safety.

Employment Relations Today
Topics: Hazardous substances, chemicals; labeling; information dissemination; safety training; communications; regulations; state laws; compliance; disabilities; injuries; industrial accidents; prevention; safety management; wellness programs.

HR Magazine
Topics: Safety management; occupational safety; work environment; safety training; electronics/computer industry; case studies.

Industrial Management
Topics: Safety training; safety program; industrial plants; contractors; accident prevention.

International Labor Review
Topics: Occupational safety in EC and Third World.

Modern Materials Handling
Topics: Occupational safety; materials and handling; injuries; ergonomics; workers compensation; OSHA; working conditions; labor force; productivity.

Occupational Hazards
Topics: Safety training; safety management; audits; case studies; industrial accidents; accident prevention; OSHA.

Occupational Safety and Health
Topics: Safety management; safety training.

Personnel
Topics: Safety management; safety training; electronics and computer industry; stress accidents.

Personnel Administrator
Topics: Safety management; occupational safety; safety program; incentives; behavior modification.

Personnel Management
Topics: Occupational safety in U.K. and E.C.; federal legislation; requirements; reforms.

Professional Safety
Topics: Occupational hazards; safety management; safety training; industrial safety; accidents; negligence; safety environmental integration; federal legislation; compliance; management roles and responsibilities.

Safety Management
Topics: Occupational hazards; industrial accidents; safety programs; safety training; risk management.

Training and Development Journal
Topics: Accident prevention; safety training; supervision; regulations; OSHA.

Training
Topics: Mandatory training; regulated industries; airline industry; mass transit; nuclear power plants; case studies.

Glossary

OSHA HEALTH HAZARDS

Agents acting on blood hematopoietic system   Chemicals that decrease hemoglobin function, depriving the body tissues of oxygen. Example: Carbon monoxide, cyanides.

Agents damaging the lungs   Chemicals that irritate or damage the pulmonary tissues. Example: silica, asbestos.

Carcinogen   A substance capable of causing or producing cancer in mammals, including humans. Example: tobacco, benzene.

Cutaneous hazards   Chemicals that effect the dermal layer of the body causing defatting of the skin, rashes, and irritation. Example: ketones, chlorinated compounds.

Eye hazards   Chemicals that effect the eye or visual capacity, causing conjunctivitis and corneal damage. Example: organic solvents, acids.

Hepatotoxin   Chemicals that produce damage indicated by jaundice and liver enlargement. Example: Carbon tetrachloride, nitrosamines.

Irritant   Chemicals that cause reversible damage to skin or mucous membranes. Example: mineral spirits.

Nephrotoxin   Chemicals that produce kidney damage. Example: halogenated hydrocarbons, uranium.

Neurotoxin   Chemicals that have their primary toxic effect on the nervous system indicated by narcosis or behavioral changes. Example: mercury, carbon disulfide.

Reproductive toxin   Chemicals that effect reproductive capabilities, including chromosomal damage (mutations) and effects on fetuses (teratogenesis). Symptoms include birth defects, sterility. Example: lead, DBCP.

Sensitizer   Chemicals that cause an allergic reaction in a substantial number of exposed people. Example: Chlorinated solvents.

Toxic or highly toxic agent   Chemicals that cause central nervous system or respiratory damage, unconsciousness, or death. Example: organic solvents, carbon monoxide.

OSHA PHYSICAL HAZARDS

Combustible liquid   Any liquid having a flashpoint at or above 100°F and below 200°F.

Compressed gas   A gas or mixture of gases having, in a container, an absolute pressure: (a) exceeding 40 psi at 70°F, or (b) exceeding 104 psi at 130°F. Or a liquid having a vapor pressure exceeding 40 psi at 100°F.

Corrosive   A chemical that causes visible destruction or irreversible alterations in living tissue by chemical action at the site of contact. Example: sodium hydroxide, sulfuric acid.

Explosive   A chemical that causes a sudden, almost instantaneous release of pressure, gas, and heat when subjected to sudden shock, pressure, or high temperature.

Flammable aerosol   A suspension of liquid or solid particles in a gas that when ignited yields a flame projection exceeding 18 inches at full value opening or a flashback (flame extending back to the valve) at any degree of valve opening.

Flammable gas   A gas that at ambient temperature and pressure forms a flammable mixture with air at a concentration of 13 percent by volume or less, *or* a gas that, at ambient temperature and pressure, forms a range of flammable mixtures with air wider than 12 percent by volume, regardless of the lower limit.

Flammable liquid   Any liquid having a flashpoint below 100°F.

Flammable solid (DOT usage)   Any solid material, other than one classified as an explosive, that under conditions normally incident to storage is liable to cause fire through friction or retained heat from manufacturing or processing; or that can be ignited readily, and when ignited burns so vigorously and persistently as to create a serious storage hazard.

Organic peroxide   An organic compound that contains the bivalent -0-0- structure and that may be considered a structural derivative of hydrogen peroxide, where one or both of the hydrogen atoms has been replaced by an organic radical.

Oxidizer   A chemical other than a blasting agent or explosive that initiates or promotes combustion in other materials causing fire either by itself or through the release of oxygen or other gases.

Oxidizing agent   A chemical or substance that brings about an oxidation reaction. The agent may either provide the oxygen to the substance being oxidized (Example: oxygen) or receive electrons transferred from the substance being oxidized (Example: chlorine).

Pyrophoric   A chemical that will ignite spontaneously in air at a temperature of 130°F (54.4°C) or below.

Unstable/reactive   A chemical that, as produced or transported, will vigorously polymerize, decompose, condense, or become self-reactive under conditions of shock, pressure, or temperature. Example: acetylene.

Water reactive   A chemical that reacts with water to release a gas that either is flammable or presents a health hazard. Example: sulfuric acid.

WORDS ON MATERIAL SAFETY DATA SHEETS

Acid A corrosive substance with a pH below 7. Generally non-flammable. Acids cause pain on contact with human skin. Sulfuric, hydrochloric, and nitric acids are common examples. RCRA defines as corrosive an acid with a pH below 2.

Acute effect A symptom experienced during or following a single exposure to a hazardous material. An acute effect of overexposure is a short-term effect.

Asphyxiant A vapor or gas that can cause suffocation when breathed.

Base A corrosive substance with a pH above 7. Bases do not cause immediate pain sensation, thereby allowing for longer exposure. Sodium and potassium hydroxide are common examples. RCRA defines as corrosive a base with a pH above 12.5.

Boiling point The temperature at which the vapor pressure of a liquid exceeds atmospheric pressure. Low boiling liquids are more hazardous because they evaporate more quickly.

C.A.S. number An identification number assigned to a chemical by the Chemical Abstract Service.

Caustic A basic or alkaline substance (see Base).

Central nervous system The brain and spinal cord. CNS depression is characterized by alterations in consciousness, from drowsiness or drunkenness to coma.

Chronic effect Illness or disease that develops slowly over a long period of time or upon repeated, prolonged exposure to a hazardous material.

Cryogen A gas that must be cooled to less than $-150°F$ to be liquified.

Decomposition The breakdown of a chemical by heat or chemical reaction.

Defatting The removal of natural oils from the skin by a fat-dissolving solvent.

Dermatitis Inflammation of the skin, characterized by oozing, redness, and itching.

Evaporation rate The rate at which a liquid will turn into a vapor. The lower the boiling point, the faster the evaporation rate.

Flammable limits The range of a vapor concentration in air that will burn or explode if an ignition source is present. The lower explosive limit (LEL) is the percentage of a material in air below which the concentration is too lean to burn. The upper explosive limit (UEL) is the percentage above which the concentration is too rich to burn.

Flashpoint The minimum temperature at which a liquid gives off enough vapor to form an ignitable mixture with air.

Hazardous chemical Any chemical for which there exists a health hazard or physical hazard.

Ignition source Anything that provides heat, spark, or flame sufficient to cause combustion or explosion.

Intoxication A toxic reaction or sign of overexposure to a chemical.

Mist Liquid particles suspended in air.

Non-combustible A material that will not ignite, burn, support combustion, or release flammable vapors when exposed to heat or fire.

PEL (permissible exposure limit) An employee exposure limit set by OSHA.

Glossary 223

Physical hazard A chemical that is a combustible liquid, a compressed gas, explosive, flammable, an organic peroxide, pyrophoric, unstable/reactive, or water-reactive.

ppm (parts per million) A unit for measuring the concentration of a gas or vapor in a million parts of air.

Respiratory system The breathing system, including the lungs, windpipe, larynx, mouth, and nose.

SCBA A self-contained breathing apparatus, which is a respirator connected to an air source.

Simple asphyxiant A chemical that causes suffocation by displacing oxygen. Acetylene, argon, and nitrogen are common examples.

Sublime To pass directly from solid to vapor form. Acrylamide is a substance that sublimes.

Symptom An abnormality in the body that is noticeable to the person experiencing it and may indicate poisoning or disease.

TLV (threshold limit value) A term used to express the airborne concentration of a material to which nearly all persons can be exposed day after day without acute or chronic effects.

Toxin Poison.

Toxic reaction A symptom or symptoms felt by a person during or after initial or repeated exposure to a hazardous material. Symptoms range from skin, eye, and nose irritation to dizziness, difficulty breathing, and unconsciousness.

Vapor The gaseous form of a substance.

Vapor density The relative weight of a vapor or gas (with no air present) as compared with an equal volume of air at ambient temperature. Chemicals are said to be lighter or heavier than air.

Vapor pressure The pressure of the vapor in equilibrium with the liquid at the specified temperature. Higher values indicate a higher evaporation rate.

Index

Accidents:
    causes of, 189-190
Acute effects, 204
Acid, 190
Air and Waste Management
    Association, 211
Air pollution:
    permit, 14
    prevention of, 32, 72, 205
Alarm systems:
    OSHA requirements for (29 CFR
        1910.165), 86
    RCRA requirements for, 88
    use in emergencies, 86-88
    See also Contingency Plan;
        Emergency; Emergency and
        Fire Prevention Plan;
        Preparedness and Prevention
        Plan
American Institute of Hazardous
    Materials, 211
American Society of Safety
    Engineers, 211
American Society of Testing and
    Materials (ASTM), 211
    standard guide for RCRA training,
        85

Audit:
    housekeeping, 149-160
    inventory control, 153-156
    materials storage, 150-153, Figure
        7-1
    spill/leak control, 158-160
    waste management, 156-159,
        Figure 7-2
    See also Housekeeping;
        Housekeeping procedures

Base, 190
Business manager:
    role of, general, 22, 163-164

Carcinogens, 33-34, 218
Car dealership. *See* Vehicle repair
Change:
    in company management, 140
    in employee behavior, 24-25,
        144-145, 182, 196, 205
    in operating procedures, 24-25,
        143
Chemical exposure:
    employee, levels of, 31-32, 37-40,
        Figure 3-1

Chronic effects, 204
Commercial printing. *See* Printing industry
Communication:
  among managers, 21, 161-168
  barriers to effective, 134, 148, 187-190
  between employees and management, 17, 20-21, 28, 134, 141-143, 186-187
  listening as a key part of, 148, 183, 192-195
  role of in compliance, 20-21, 146, 148, 167
  trust as a key part of, 17, 28, 184, 193
  ways to improve, 20, 148, 195
Communications systems:
  OSHA requirements for (29 CFR 1910.165), 86
  RCRA requirements for, 88
  use in emergencies, 86-88
  See also Contingency Plan; Emergency; Emergency and Fire Prevention Plan; Preparedness and Prevention Plan
Compliance, 2-4;
  as a production concern, 142, 169
  assessing success of, 16, 144-145
  benefits of, 2, 166, 169
  coordinating efforts toward, 169
  employee involvement in, 12, 145, 149, 181
  meeting workplace needs through, 3-5
  myths regarding, 142-145, 162-163
  role of emergency planning in, 84-85
  role of housekeeping in, 160
  role of management in, 161-168, 177
  role of materials storage in, 151, 153
  role of protective equipment in, 83-84
  role of training in, 161, 163-164, 172
  See also Communication; Goal-setting; Training regulations
Compressed gases, 107, 189, 192
Confined space entry, 180
Contingency Plan (RCRA 40 CFR 264):
  evacuation plan requirements of, 91
  general requirements of, 84, 88, 171, 181
  intersection with OSHA Hazard Communication Standard, 173-174
  intersection with OSHA Lockout/Tagout Rule,
  intersection with OSHA Emergency and Fire Prevention Plan, 84-85, 88-89, 91, 93, 99, 105-108, 175
  See also Alarm systems; Communications systems; Emergency Coordinator; Evacuation; Hazardous waste; Hazardous waste generator; Preparedness and Prevention Plan
Corrosive, 38-40, 190;
  handling of, 189
  storage of, 158
CPR, 92-94, 100
Cutting. *See* Welding

Department of Transportation (DOT), 84, 168;
  See also Director of Transportation; Truck driver

Index   227

Director of Transportation:
   role of, at TSDF, 168

Egress:
   means of (OSHA 29 CFR 1910.37), 102
   See also Exits
Electroplating. *See* Plating industry
Emergency:
   management, 85, 93
   methods of reporting an, 86
   OSHA definition of an, 86
   preparation for an, 83
   prevention of an, 83, 98-99, 151, 174-175
   response assistance, 212-213, 215
   training, 96-101, 174-175
   See also Emergency Action Plan;
      Emergency Action Plan
      training; Emergency and Fire
      Prevention Plan; Evacuation;
      Fire Prevention Plan;
      Protective equipment
Emergency Action Plan (OSHA 29 CFR 1910.38):
   communications and alarm systems in, 86-88, 94-95
   content of, 86-95
   development, employee involvement in, 97-100
   evacuation procedures in, 88-91, 94-95
   leadership roles in, 92-95, 100
   purpose of, 86, 99-100
   rescue procedures in, 92, 94
   shutdown procedures in, 91, 94, 100
   See also Alarm systems;
      Communications systems;
      Contingency Plan;
      Emergency; Emergency
      Coordinator; Emergency and
      Fire Prevention Plan;
      Evacuation; Fire Prevention
      Plan; Preparedness and
      Prevention Plan; Resource
      Conservation and Recovery
      Act
Emergency Action Plan training, 99-101;
   requirements in Emergency and Fire Prevention Plan Standard, 95-96
   using company meetings to meet requirements, 101
   using drills in, 101-102
   See also Emergency Action Plan
Emergency action team:
   description of, 93-95
   roles of, 93-95, Figure 5-1
   training of, 100
   See also Emergency Action Plan
Emergency and Fire Prevention Plan (OSHA 29 CFR 1910.38):
   intersection with OSHA Hazard Communication Standard, 174-176, 84
   intersection with OSHA Lockout/Tagout Rule, 110
   intersection with OSHA personal protective equipment standards, 84
   intersection with RCRA Hazardous Waste Generator requirements, 174-176
   general requirements of, 83-85, 151, 171, 174, 176
   training, content of, 174-176
   See also Alarm systems;
      Emergency Action Plan;
      Emergency; Communications
      systems; Evacuation; Fire
      Prevention Plan
Emergency coordinator:
   RCRA requirement for, 89, 92

228   Index

Emergency coordinator *(Continued)*
  role of, 92-93, 95
  selection of, 92-94
  training of, 96-100
  See also Contingency Plan
Emergency response workers:
  training for, 8, 178-180
  See also Hazardous Waste
    Operations and Emergency
    Response
Employee(s):
  changing behavior in, 32, 144-145,
    182, 196, 205
  encouraging innovative thinking in,
    2, 157
  lack of care on behalf of, 64, 202
  illiteracy in, 187-189
  older, 188-189
  See also Communication;
    Compliance
Empty container:
  definition of, 157
  management of, 157-158
Energy hazards:
  definition of, 109, 113
  identification of, 116-117, 123
  minimization of, 147
Energy isolating devices:
  chart of, 119, 123
  identification of, 118-120
  labeling of, 120
Energy sources:
  definition of, 109
  isolating machines from, 118-125
  stored energy, release of, 124-125
  identification of, 119
Engineering controls, 63
Environment:
  impact of individuals on, 8-9, 12,
    14, 205
  impact of waste on, 14, 200
  workplace impact of on, 1, 172, 205

Environmental manager:
  role of, at manufacturing company,
    166
  role of, at TSDF, 162-163, 167-168
  role of, general, 19, 161-162,
    163-165
Environmental Protection Agency
  (EPA), 144-145, 150, 169;
  permit requirements, 161-163
Environmental training:
  coordinating compliance through,
    163-164, 1647-168, 172-177
  documentation of, 177-179, Figure
    8-2
  role of managers in, 161-165,
    167-168
  See also Safety training
Evacuation:
  drills, 89, 101-102
  equipment shutdown during, 91
  exits, 102-103
  leadership during, 89, 93, 97-98
  maps, 88-89, 90
  plan, 89-91, 99-100
  rally point, 89-90
  roll call after, 89
  routes, 88, 98
  supervisor roles, 93, 94, 97-98
  training, 90, 97-101, 177
  See also Contingency Plan;
    Emergency Action Plan,
    Emergency and Fire
    Prevention Plan
Exits:
  inspection of, 102-103
  labeling of, 103
  See also Egress
Eye protection:
  availability of, 67-68
  chemical splash goggles, 67-70
  comfort in, 69-70
  company policies regarding, 68-70

Index 229

face shield, 67
OSHA requirements (29 CFR 1910.133), 67
reason for use, 67
safety glasses, 66-70
selection of, 68
use of, 146

Fire hazards:
  identification of, 97-98, 104-105
  minimization of, 32, 107-108, 147, 151
Fire Prevention Plan (29 CFR 1910.38):
  development, employee involvement in, 105, 108
  fire control equipment, definition of, 105
  fire control equipment, maintenance of, 106
  fire prevention equipment, definition of, 105
  fire prevention equipment, maintenance of, 105-107
  fire prevention procedures, 104-105, 107
  fuel sources, control of, 107-108, 151
  fuel sources, definition of, 106-107
  general requirements of, 103-104
  intersection with OSHA Hazard Communication Standard, 107-108
  intersection with OSHA Lockout/Tagout Rule, 110
  purpose of, 104
  using housekeeping procedures, 107-108, 151
  See also Emergency Action Plan; Emergency and Fire Prevention Plan; Housekeeping; Preparedness and Prevention Plan
Fire Prevention Plan training, 105-108
First Aid, 92-94, 100
Flammables, 32, 39-40, 47;
  as fuel sources, 107
  fire hazards of, 104-105, 107
  fire involving, 97-98
  storage of, 150, 158, 174, 192, 201
Flashpoint, 32, 40

Games:
  use in training, 200-201
Gloves, 64, 66;
  use with solvents, 141
Goal-setting:
  for improving the workplace, 146-148
  in training programs, 21-22, 25, 27, 161, 177, 183, 198-199
  role in compliance process, 143-148, 160, 161, 167-170, 172, 180

Hazard Communication Standard (OSHA 29, CFR 1910.1200), 1;
  general requirements of, 40-58, 154, 170
  intersection with OSHA Emergency and Fire prevention Plan, 174-175
  intersection with OSHA Lockout/Tagout Rule, 110
  intersection with RCRA Hazardous Waste Generator requirements, 172-174, 178
  purpose of, 6, 35, 144-145
  role of protective equipment in, 63-64
  written program, 42-44, 47, 144
  See also Hazardous materials; Health hazards; Labeling;

Hazard Communication Standard
 (*Continued*)
  Material Safety Data Sheets;
   Physical hazards; Protective
   equipment; Right-to-Know
Hazard Communication training:
  content, 63, 66, 188-189, 198-200
  requirements in Hazard
   Communication Standard,
   56-58, 63-64, 196, 54,
   172,-174, 176, 180-181
  See also Right-to-Know
Hazard prevention, 72
Hazardous materials:
  assessing employee interaction
   with, 37-40
  behavior of, 31-32
  fire hazards of, 104-105
  OSHA definition of, 33-34
  protecting employees from, 63
  management of, 151
  resources regarding, 214, 218
  storage and handling of, 104-105,
   150-151, 174, 201
  See also Labeling; Hazard
   Communication Standard,
   Material Safety Data Sheet;
   Protective equipment
Hazardous waste:
  definition of, 9, 173
  inspection, 106
  management, RCRA requirement
   (40 CFR 262), 156, 171,
   205
  storage/handling of, 97-98, 104,
   106, 150-151, 156, 174
  See also Hazardous waste
   generator; Labeling; Resource
   Conservation and Recovery
   Act
Hazardous waste generator:
  intersection with OSHA Emergency
   and Fire Prevention Plan,
   174-175
  intersection with OSHA Hazard
   Communication Standard,
   172-174, 178
  RCRA requirements, general, 13,
   84-85, 106, 108, 151, 156
  RCRA requirements, training, 7-8,
   13, 84, 173-174, 176, 181
  See also Contingency Plan;
   Hazardous waste; Resource
   Conservation and Recovery
   Act
Hazardous waste management
  company. *See* Treatment,
  Storage, and Disposal Facility
Hazardous Waste Operations and
  Emergency Response (OSHA 29
  CFR 1910.120):
  purpose of, 8-9
  requirements of, 83-84, 180
Hazardous Waste Operations and
  Emergency Response training,
  180
Health hazards:
  definition of, 34, 194,198-199
  minimization of, 32, 147
  training in, 63, 173, 176, 194,
   198-199
Hearing Conservation Standard
  (OSHA 29 CFR 1910.95), 82
Hospitality industry:
  housekeeper, protective equipment
   use by, 67-68
  labeling for, 46-47
  regulations affecting, 46-47, 67-68
Housekeeping:
  definition, 149
  role in compliance, 149, 160
  See also Audit; Housekeeping
   procedures
Housekeeping procedures:

Index 231

for inventory control, 153-156, 160
for management of empties, 158
for materials storage, 150-153, 160
for spills and leaks, 158, 160
for waste management, 156-158, 159-160
for solvent wipes and paint filters, 158
to prevent fire, 107-108, 149, 151, 158
See also Audit; Housekeeping
Human resources manager:
  role of, at manufacturing company, 165-167
  role of, general, 57, 163-164

Ignition sources, 104-105
Incentives, 22
Inspection:
  of employee safety behavior, 146
  of fire prevention equipment, 106
  of protective equipment, 66, 180
  of respirators, 77-79, 80, Figure 4-2
  of workplace by OSHA and EPA officials, 144-145, 150
  of workplace for safety and environmental compliance, 162-163
  See also Audit; Lockout/Tagout Rule; Protective equipment; Respirator
International Fire Service Training Association, 211-212
Inventory control:
  benefits of, 156
  procedures for, 153-155
  See also Housekeeping procedures; Safe operating procedures; Waste minimization

Job description, 36;
  housekeeping as part of every, 160

Job pride, 11

Labeling:
  Hazard Communication requirement for, 44-47
  hazardous materials, 44-47, 180, 192, 201, 214
  hazardous waste, 156, 174, 180
  portable container, 47-48
Laws:
  attitudes toward, 3, 5
Leaks. *See* Spills
Lockout:
  definition of, 121
  release from, 125-126
Lockout/Tagout Rule (OSHA 29 CFR 1910.147):
  affected employee, 116, 122-128, 131-132
  authorized employee, 116, 122-128, 130-131, 134
  energy control procedure, elements of, Figure 6-1, 117
  energy control procedure requirements of, 115-130
  energy control policy, 114-116, 135
  energy control program, requirements, 113-114, 126
  equipment isolation, verification of, 124-125
  equipment shutdown, 120-121
  exception to requirement for written procedure, 136-137
  exemptions, 112-113
  general requirements of Rule, 111-112, 171
  group lockout/tagout, 126-127
  inspections, 128-129, 132-134
  intersection with OSHA Emergency and Fire Prevention Plans, 110
  intersection with OSHA Hazard Communication Standard, 110

Lockout/Tagout Rule *(Continued)*
  purpose of, 7, 109, 113, 134, 144-145
  routine and non-routine maintenance, definition, 117-118
  stored energy, release of, 124-125
  See also Energy isolating devices; Lockout; Lockout/Tagout training; Locks; Tags; Tagout
Lockout/Tagout training:
  contract employees, 135-136
  documentation of, 132-134, Figure 6-4
  frequency of, 132-134
  program, content of, 110, 119, 124, 130-132, 134
  program, purpose of, 131-132
  requirements in Lockout/Tagout Rule, 129-130
Locks:
  removal of, 126-128
  use of, 122-124, 145

Machinery and machine guarding (OSHA 29 CFR Subpart O):
  requirements of, 112
  standard operating procedures for, 109
  See also Lockout/Tagout Rule
Machining industry:
  Lockout/Tagout program for, 123, Figure 6-3
  Right-to-Know training for, 185, 190-191
Maintenance:
  definition of, 110
  energizing equipment during, 122
  non-routine, 118, 139
  preventive, 106-107, 137-139
  protection during, 112-113, 117-118, 138

  role of lockout/tagout in, 109, 112-113, 137-139
  routine, 117-118, 137-139
Management:
  involvement in compliance, 1, 23-25, 57, 164-168, 177
  involvement in safety, 1, 60, 71
Manufacturing industry:
  communications and alarm systems in, 86-87
  compliance in, 165-168
  emergency training in, 97-98
  painting department, standard operating procedures of, 141
  protective equipment use in, 70-71, 146-148
  respirator training in, 165
  Right-to-Know training in, 165-167
  secretary, training needs of, 37-38
Material Safety Data Sheet (MSDS):
  exemptions for, 51
  group, 52-56, Figures 3-3, 3-4, 3-5
  Hazard Communication requirements for, 48-52
  source of emergency information, 90-91
  source of protective equipment information, 68, 73-74
  usefulness of, 51-52
  use in training, 199-200
  See also Hazard Communication Standard
Mechanical plating. *See* Plating industry
Morale, 11, 64, 70

National Environmental Training Association, 212
National Fire Protection Association, 212
National Safety Council, 212

Occupational Safety and Health Administration (OSHA), 143, 150, 158, 161, 165, 168, 194; Standards, performance-oriented, 66, 144-145, 174
Overexposure:
  to chemical vapors, 32, 40, 54, 190, 199

Paint filters:
  waste, storage of, 105, 150, 158
Permissible exposure limit (PEL), 194
Physical hazards:
  definition of, 34, 107
  minimization of, 147
Plant manager:
  role of, at manufacturing company, 147
  role of, at TSDF, 168
Plating industry:
  chemical handling in, 13
  dock worker, training needs of, 38-39
  maintenance employees in, 11, 91
  Right-to-Know training in, 11-12, 38-39, 185
  spill management in, 13, 90-91
  wastewater treatment in, 12-13
Pollution prevention, 2;
  plan, 14
  regulatory requirements for, 151, 157
  role of air pollution prevention in, 151, 156, 160
  role of materials storage in, 151, 155-156
  role of spill prevention in, 151, 156, 158, 160
  See also Waste minimization
Preparedness and Prevention Plan (RCRA 40 CFR 264):
  intersection with OSHA Emergency and Fire Prevention Plan, 84-85, 88, 99, 106, 175
  intersection with OSHA Hazard Communication Standard, 173-174
  requirements of, 84, 171
  See also Alarm systems; Communications systems; Contingency Plan; Egress; Evacuation; Exits; Hazardous waste
Printing industry:
  film processor, training needs of, 39
  Lockout/Tagout program for, 118-120
  press operator, protective equipment use by, 63-64
  press operator, training of, 185
  regulations affecting, 18-19, 170-171
Production manager:
  perspective on training, 57, 186-187
  role of, at manufacturing company, 167
  role of, at printing company, 168-170
  role of, general, 163-164
Production operations:
  efficiency in, 13
  improving, 10
  streamlining, 10
Protective clothing, 64-66;
  See also Eye protection; Gloves; Protective equipment; Respirator
Protective equipment:
  definition of, 65
  guidelines for selection of, 65, 181, 213-214
  intersection with other regulations, 147, 175-177
  maintenance and cleaning of, 66, 69

Protective equipment *(Continued)*
  OSHA standards, (29 CFR 1910.132, 133, 134) general requirements of, 64-66, 83, 170-171
  policy for use of, 11, 63-64
  program, management involvement in, 71
  purpose of, 63-64, 83-84
  training, 63, 96, 180-181
  types of, 63-64
  use in emergencies, 77, 83
  use of, 65-66, 137-138, 146-147, 177, 179-180, 205
  See also Eye protection; Gloves; Protective clothing; Respirator
Purchasing manager:
  with safety responsibilities at manufacturing company, 70-71
  with safety responsibilities at TSDF, 162-163, 167-168

Quality, 1, 11, 26, 187, 205; of life, 31, 33, 35

Rags. *See* Solvent wipes
Regulations:
  as context for training, 176
  employee awareness of, 11, 14
Repair. *See* Maintenance
Resource Conservation and Recovery Act (RCRA), 1, 158;
  Personnel Training requirements (40 CFR 264.16), 7-8, 10, 84-85, 88, 91, 96, 105-106, 108, 173-174, 176, 181
  See also Hazardous waste; Hazardous waste generator; Waste minimization

Respirator:
  air purifying, 70, 72-73, 75, 80, 146-147, 180
  employee responsibility for, 73
  fit testing of, 75, 180
  maintenance and cleaning of, 66, 75-80
  repair of, 79
  SCBA, 76-77, 79, 180
  selection of, 73-74
  storage of, 80-82
  training card, 76, Figure 4-1
  training, 73, 75-77, 81
  use in IDLH atmospheres, 76, 83
  use of, 70-71, 73-76, 146-148
  See also Inspection; Respiratory Protection Standard
Respiratory health:
  shared employee/employer commitment to, 72
Respiratory Protection Standard (OSHA 29 CFR 1910.134):
  as guideline, 7
  compliance, evaluating success of, 148
  needs assessment, 73
  general requirements of, 72, 171
  written program, excerpts from, 75-76, 79-82
  written program, requirements of, 73-74
Right-to-Know:
  as foundation of all training compliance, 35, 42, 60-62
  as source of power, 29-35, 60
  intersection with other regulations, 58-60, 173-175
  as guideline for better management, 28-29, 60-62
  as guideline for meeting employee needs, 34-40
  training, content of, 5-7, 11-13,

Index   235

33-34, 56-60, 141, 172-174,
176-177, 181, 190-191
training, purpose of, 6, 31-33, 35,
144
See also Hazard Communication
Standard
Roll Call Officer:
role in evacuation, 89, 93-94
See also Emergency and Fire
Prevention Plan; Evacuation

Safe operating procedures:
definition of, 143-144
for housekeeping, 154-157
for materials storage, 151, 153-154
for materials use, 155-156
for spill and leak prevention, 160
incorporated into training, 145
practicality of, 146, 148, 153,
160
training in, 143, 147, 160, 189
See also Standard operating
procedures
Safety:
definition of, 1-2
employee attitudes toward,
140-141, 177, 181
meetings, as opportunity to meet
training requirements, 178-181
meetings, employee involvement
in, 181
See also Safe operating procedures;
Standard operating procedures
Safety manager:
role of, at manufacturing company,
97-99, 146-148, 166
role of, at TSDF, 168
role of, general, 19, 57, 66-67,
161-165
Safety training:
content of, 2, 143

coordinating compliance through,
161-164, 1647-168, 172-177
definition of, 1-2
documentation of, 177-179, Figure
8-2
role of managers in, 161-165,
167-168
using safety meetings in, 70, 79,
178-181
See also Environmental training
SARA Title III, 14; Form R, 14
Servicing. *See* Maintenance
Solvent wipes:
storage of, 105, 150, 158
Spill:
clean-up, 192, 205
negative impacts of, 160
protective equipment use during
clean-up of, 178, 195
sources of, 158, 160
training in, 173-174, 176-177,
178-181
See also Housekeeping procedures;
Pollution prevention; Safe
operating procedures; Waste
minimization
Standard operating procedures:
as basis of training, 57, 140
definition of, 141-142
emergency prevention, 98-99,
138-139
emergency response, 119
energy control, 109, 134, 137-139
incorporating safety and
environmental concerns,
140-142, 148
machine guarding, 109-138
maintenance, 119, 135, 137-139
respirators, 73
See also Safe operating procedures
Superfund:
site, clean-up at, 8

Tagout:
    definition of, 121-122
    release from, 125-126
Tags:
    removal of, 126-128
    use of, 122-124, 145
Threshold limit value (TLV), 32, 194, 218
Toxin, 194
Trainers:
    resources for, 209-211
    training for, 215-218
Training programs:
    assessing employee needs under, 40
    benefits of, 10
    communication in, 20-21
    consistency in, 16-20
    discipline in, 23-25
    documentation of, 177-178, Figure 8-2
    enforcement of, 19-20
    leadership in, 26-27
    purpose of, 2-4, 172
    support of, 19
    See also Change; Compliance; Environmental training; Safety training; Training regulations
Training regulations:
    as guidelines for meeting needs, 5-9, 42, 64, 145, 165
    as tools for changing behavior, 63-64, 143, 171-172, 174
    benefits of, 8, 10-15, 145-166
    defense of, 6, 9
    intersections among, 172-177
    purpose of, 2-5, 42
Training techniques, 182-184;
    avoiding boring presentations, 194, 200-202
    for new trainers, 186
    importance of follow-through, 193
    importance of honesty, 185

Treatment, Storage, and Disposal Facility (TSDF):
    compliance at, 84, 162-163, 168
    employee training at, 8
    materials handlers at, 10
    RCRA requirements for, 84, 162
Truck driver:
    DOT training for, 84
    respirator storage by, 81
    See also Department of Transportation

Vehicle repair industry:
    hazard prevention in, 72
    painter, training needs of, 39-40, 185
    personal protective equipment use in, 72, 80-82, 197
    Right-to-Know training for, 14, 196-197
    waste minimization in, 14
Videos, 144-145
Volatile organic compound, 12, 14

Waste management, 149-152, 156-158;
    See also Hazardous materials; Hazardous waste, Housekeeping procedures; Waste minimization
Waste minimization, 2;
    benefits of, 13-14, 151, 153-154, 156-157
    employee involvement in, 157
    maintenance and housekeeping in, 13
    plan, 14
    regulatory requirements of, 157, 171
    role of inventory control in, 154-156
    role of materials storage and use in, 154-156